Wonders of
Water

The Hydrogen Bond in Action

Wonders of
Water

The Hydrogen Bond in Action

Ivar Olovsson

University of Uppsala, Sweden

 World Scientific

NEW JERSEY · LONDON · SINGAPORE · BEIJING · SHANGHAI · HONG KONG · TAIPEI · CHENNAI · TOKYO

Published by

World Scientific Publishing Co. Pte. Ltd.

5 Toh Tuck Link, Singapore 596224

USA office: 27 Warren Street, Suite 401-402, Hackensack, NJ 07601

UK office: 57 Shelton Street, Covent Garden, London WC2H 9HE

British Library Cataloguing-in-Publication Data
A catalogue record for this book is available from the British Library.

WONDERS OF WATER
The Hydrogen Bond in Action

ISBN 978-981-3229-11-2
ISBN 978-981-3235-16-8 (pbk)

Preface

Water plays a unique role in chemistry. The special properties of the different forms of water — from ice and snow to liquid water — are due to hydrogen bonding between the H_2O molecules, and this book is a tribute to the hydrogen bond, a field to which I have spent a major part of my research. The hydrogen bond is of fundamental importance in biological systems since all living matter has evolved from and exists in an aqueous environment. Hydrogen bonds are involved in most biological processes as little energy is needed in forming as well as breaking of these bonds. Without hydrogen bonds, no water can be transported from the roots to the leaves in the trees!

According to the classical Chinese text *Tao Te Ching*, "The highest excellence is like water. There is nothing in the world softer and weaker than water, and yet, when it comes to attacking things that are firm and strong there is nothing that can surpass it — because there is nothing that is so effective that it can replace water." This may look like a strong exaggeration but it is very true: All over the world you will find a countless number of traces from the strong action of water and ice through the history of the Earth. A few examples are shown in the book.

The present volume is a considerable extension of my previous book *Snow, Ice and Other Wonders of Water*. Liquid water and ice are more than ever objects of intensive research and it would have been tempting to include much more of the large material which has appeared in recent years. However, the purpose of this book is just to whet the appetite for this fantastically interesting field. It has therefore been necessary to leave out results which are somewhat difficult to inform about, using a simple language. I still hope that the new material will make you even more tempted to learn everything about water.

I wish to express my sincere gratitude to the Centre for Ice and Climate at the Niels Bohr Institute, University of Copenhagen and to the Electron and Confocal Microscopy Laboratory, Agricultural Research Service, U.S. Department of Agriculture, for permission to include a series of pictures from these institutions. Many thanks to Prof. Martin Chaplin for permission to include some pictures from his extensive site Water Structure and Science, to Prof. Kenneth Libbrecht for permission to use his snow crystal morphology diagram, and to Prof. Yoshinori Furukawa at the Institute of Low Temperature Science, Hokkaido University, the British artist Simon Beck, the Canadian photographer Don Komarechka, and the

Russian photographers Alexey Kljatov and Andrei Osokin, for kind permission to include their beautiful and sometimes unique pictures. I hope the pictures reproduced here will stimulate further studies of their books.

Many thanks to Professors Stefan Bengtsson, Anders Eriksson, Anders Liljas and Erling Ögren for checking parts of the new manuscript and for valuable suggestions.

Ivar Olovsson
Uppsala, 15 October 2017
Ivar.Olovsson@gmail.com

Contents

1

There are Many Different Types of Snow

Frozen water — snow and ice — appears in a myriad of different shapes and with properties which can be quite different. Detailed knowledge of the properties of snow is of great importance for the Sami people (Laplanders) involved in reindeer herding. A large number of names are used by the Laplanders to characterize the different types, and snow may still be a daily topic of conversation. In Yngve Ryd's book (in Swedish) *Snö: En Renskötare Berättar* (*Snow: A Laplander Narrates*), more than 300 words for "snow" are documented with explanations and photos. The words describe for instance the amount of snow, consistency, gliding, buoyancy or melting.

In this book I will mostly use the term "snow crystal" for a single crystal, i.e. a sample which is continuous and has no boundaries (all parts of the crystal are extinguished simultaneously in polarized light). A fully developed dendritic snow crystal is, for example, a single crystal. All snow crystals (in my terminology) are transparent. As a snow crystal falls toward the earth, it will often hook onto

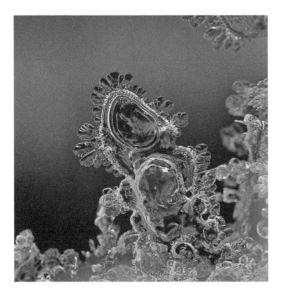

other crystals in a random way and a *snowflake* is formed. A layer of snow looks white owing to repeated reflection of the light toward the randomly oriented snowflakes. In the literature the word "snowflake" seems to be used for all types — a single crystal as well as a random collection of snow crystals. A snow crystal is just ordinary ice, but ice with a special, mostly rather open structure. A large, more compact and irregular crystal is better named *ice crystal* (see photo on the left by Andrei Osokin).

2

Early Snow Crystal Observations

Snow has always fascinated mankind and is mentioned in a dozen places in the Bible. In Job 38:22–23 is written: "Hast thou entered into the treasure house of the snow, or hast thou seen the treasure house of the hail, which I have reserved against the time of trouble, against the day of battle and war?" Hail is often considered to be punishment from God — in Revelation 16:21 is written: "And there fell upon men a great hail out of heaven, every stone about the weight of a talent: and men blasphemed God because of the plague of the hail; for the plague thereof was exceeding great." (One talent was about 50 kg.)

To my knowledge, the first pictures of snow crystals were published in 1555 by Olaus Magnus in his famous *Historia de Gentibus Septentrionalibus* (*History of the Nordic People*, an assembly of essays in 22 volumes; Fig. 2.1). Owing to the Reformation in Sweden, he lived at that time in Rome together with his brother, Catholic archbishop Johannes Magnus. In exile he seems to have largely forgotten how snow crystals look and applied his fantasy.

It is commonly considered that it was the astronomer Johannes Kepler who, in his essay *De Nive Sexangula*, first established that snow crystals have a six-fold symmetry (Figs. 2.2 and 2.3). This conclusion appears to be based on his studies of the closest packing of spheres.

Fig. 2.1. Snow crystals drawn by Olaus Magnus.

Fig. 2.2. Johannes Kepler   **Fig. 2.3.** Kepler's book *De Nive Sexangula*
(1571–1630).   (*On the Six-Cornered Snowflake*), published in 1611.

The French philosopher and mathematician René Descartes (Fig. 2.4) made, during the unusually cold winter in Amsterdam in 1635, the first detailed observations of snow crystals, which were published in 1637 in his famous work *Discours de la Methode* (Fig. 2.5). The pictures in Fig. 2.6 are found in the chapter "*Les Meteores.*" This may seem strange in a treatise dealing with snow, but *Meteorology* deals with all atmospheric phenomena, such as wind, storms, cyclones, rain, snow and hail.

Descartes was born at La Haye in Touraine, France. The village is nowadays named "Descartes," in his honor. In 1649 he was invited to Stockholm by Queen Kristina to be her teacher and adviser, and to organize a new scientific academy

Fig. 2.4. René Descartes (1596–1650).   **Fig. 2.5.** *Discours de la Methode* (1637).

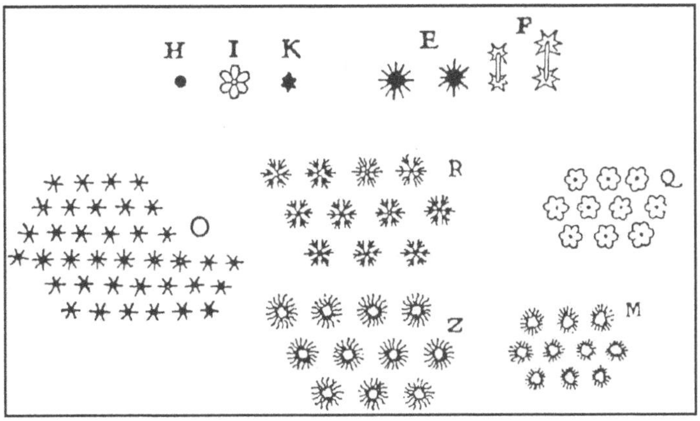

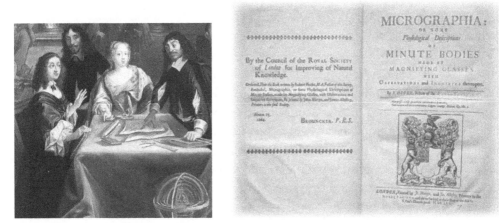

Fig. 2.6. Snow crystals observed by Descartes.

Fig. 2.7. Descartes and Queen Kristina.

By the Council of the ROYAL SOCIETY of London for Improving of Natural Knowledge.

BROUNCKER. P.R.S.

MICROGRAPHIA:
OR SOME
Physiological Descriptions
OF
MINUTE BODIES
MADE BY
MAGNIFYING GLASSES.
WITH
OBSERVATIONS and INQUIRIES thereupon.

Fig. 2.8. Robert Hooke's *Micrographia*, published in 1665.

(Fig. 2.7). It has been said that Descartes considered Sweden a country where both people and thoughts froze to ice. The meetings were held early in the morning (at 5 a.m.) in the castle which was hardly heated and very cold and draughty. Descartes caught a cold and died of pneumonia on February 11, 1650, after only a few months in Sweden.

Many other scientists have also pondered on the mysteries of snow crystals. When the microscope was invented in the later part of 1600, the possibilities of studying snow crystals became much better. In 1665 the multidimensional scientist and polymath Robert Hooke published *Micrographia: Or Some Physiological Descriptions of Minute Bodies Made by Magnifying Glasses with Observations and Inquiries Thereupon* (Fig. 2.8). An imagined picture of Hooke at his desk is shown in Fig. 2.9 (no contemporary portrait has been preserved). A few of his drawings of snow crystals are shown in Fig. 2.10. Hooke remarked that the angle between the side branches is always 60°.

Fig. 2.9. Robert Hooke (1635–1703).

Fig. 2.10. Snow crystals drawn by Robert Hooke.

The eminent natural scientist and whaler William Scoresby, Jr. (Fig. 2.11) published in 1820 a two-volume book, *An Account of the Arctic Regions with a History and Description of the Northern Whale Fishery* (Fig. 2.12). In this famous work he also made accurate observations of snow crystals, some of which are shown in Fig. 2.13. Note that Scoresby (as well as Descartes) also observed three-dimensional snow crystals: two or three planar crystals joined by an axis.

Fig. 2.11. William Scoresby, Jr. (1789–1857).

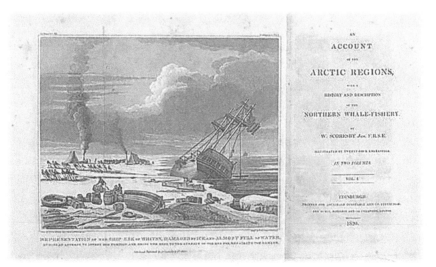

Fig. 2.12. An Account of the Arctic Regions with a History and Description of the Northern Whale Fishery.

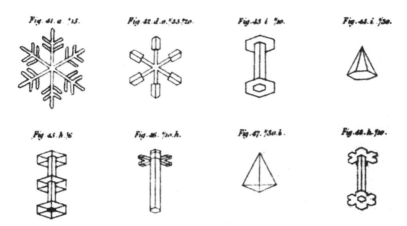

Fig. 2.13. Snow crystals observed by William Scoresby, Jr.

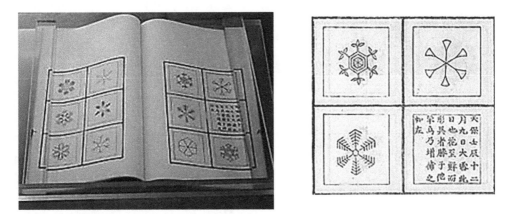

Fig. 2.14. Snow crystals from the book *Sekka Zusetsu* by Doi Toshitsura. (From an exhibit in the National Museum of Nature and Science, Tokyo, Japan.)

The *daimyo* (Japanese feudal lord) Doi Toshitsura wrote in 1832 the book *Sekka Zusetsu*, on snow crystals (Fig. 2.14). Note that in one of the pictures the branches are directed inward — a most unlikely situation, and probably never observed.

The English meteorologist and aeronaut James Glaisher (Fig. 2.15) published in 1855 a collection of snow crystals: *Photogenic Drawings of Snow Crystals, as Seen in January 1854.* The snow

Fig. 2.15. James Glaisher (1809–1903).

Fig. 2.16. Glaisher's drawings of
snow crystals.

Fig. 2.17. Wilson A. Bentley (1865–1931).

Fig. 2.18. Snow crystals by Bentley.

crystals shown in Fig. 2.16 were sketched by Glaisher and drawn properly by his wife, artist Cecilia Louisa Glaisher (1828–1892). Her drawings of snow crystals are considered to be among the best ever published.

Wilson "Snowflake" Bentley from Jericho, Vermont, USA, was one the first known photographers of snowflakes (Fig. 2.17). He was a farmer without a scientific background. He attached a bellows camera to a compound microscope and, after much experimentation, photographed his first snowflake on January 15, 1885. For almost half a century Bentley captured and photographed more than 5,000 snowflakes. He poetically described snowflakes as "tiny miracles of beauty." He wanted them to appear "like diamonds on velvet," so he carefully cut the photographs and mounted them on black paper. His pictures were spread all over the world and published in many leading magazines. In 1931 the American Meteorological Society gathered the best of his pictures in a monograph illustrated with 2,500 pictures; a few are shown in Figs. 2.18 and 2.19. Bentley died of pneumonia at his farm on December 23, 1931, after walking home six miles in a blizzard. His book *Snow Crystals* was published shortly before his death.

Fig. 2.19. Snow crystals by Bentley.

The physicist Ukichiro Nakaya also got interested in studying snow crystals and together with his students, collected a large number of snow crystals in the mountains around Sapporo, Japan. The crystals were studied in the laboratory shown in Fig. 2.20. His attempts to grow artificial snow crystals are described in next chapter.

Fig. 2.20. Nakaya studying natural snow crystals.

The crystallographer Max Perutz was also very interested in snow crystals (Fig. 2.21). It may be mentioned that in 1962 he received the Nobel Prize in Chemistry for his determination of the crystal structure of hemoglobin (after 20 years of hard, and seemingly hopeless, work). He shared the prize with John

Kendrew who succeeded in determining the crystal structure of myoglobin. These were the first crystal structure determinations of proteins and opened the large field of protein crystallography. Perutz was also actively involved in the Project Habakkuk (*cf.* Chapter 6).

Fig. 2.21. Max Perutz examining snow crystals in the cave on the Jungfraujoch glacier in 1938.

3

Artificial Snow Crystals

Wilson A. Bentley's beautiful snow pictures inspired many scientists and artists, among them the Japanese nuclear physicist Ukichiro Nakaya (Fig. 3.1). In 1932 Nakaya got a position in the newly established science faculty at the university in Sapporo on the island of Hokkaido in northern Japan. No apparatus for research in nuclear physics was available there, so he directed his interest toward research

Fig. 3.1. Ukichiro Nakaya (1900–1962).

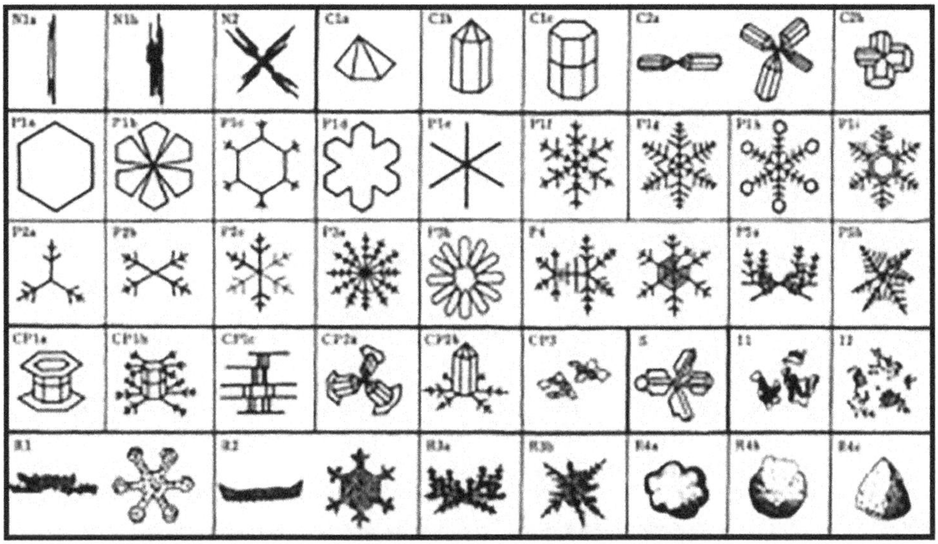

material which was unlimited around Sapporo — *snow*. During a series of winters he made careful studies of snow crystals in the mountains around Sapporo and found that regular hexagonal crystals were not as common as more irregular ones. His findings and classification are summarized in Fig. 3.2. (New classification schemes have later been introduced. The widely used Mogano–Lee scheme from 1966 contains 80 types. K. Kikuchi, T. Kameda, K. Higushi and A. Yamashita have in 2013 suggested a global classification scheme with 121 types.)

Fig. 3.2. Different snow crystal forms found by Nakaya.

Nakaya wondered why the crystals were so different — what were the atmospheric conditions when these different types of snow crystals were formed high up in the air? As it was not possible to determine the exact meteorological situation for each individual crystal, he decided to build a laboratory where he could grow snow crystals under different temperature and humidity conditions. In 1935 the low-temperature laboratory was opened. No one had previously made artificial snow crystals and it took a considerable time before Nakaya found a good method. Attempts to grow the crystals on cold surfaces only resulted in typical frost patterns. Finally, he succeeded in growing real snow crystals on rabbit hair in cold humid air, and on March 12, 1936 he created the first artificial snow crystal. Rabbit hair has small knobs at suitable distances and these are evidently suitable starting points for crystals to grow. It took 30–60 minutes before a typical branchy, dendritic snow crystal was formed.

Owing to the Second World War, it took a long time before Nakaya's collected works were published. The original text and many pictures were destroyed when the printing house was bombed. In 1954 the book *Snow Crystals: Natural and Artificial* was published. This is beautifully illustrated and summarizes Nakaya's research on snowflake crystals, starting from his work at Hokkaido University. The original book has long been out of print but reprints are available. It serves as a classic reference on crystal shapes, showing how a scientific investigation

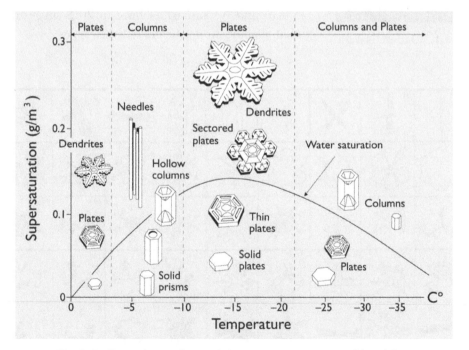

Fig. 3.3. Dependence of crystal shape on temperature and humidity.
(© Kenneth Libbrecht, permission granted.)

can proceed through systematic observation toward an accurate description of a fascinating natural phenomenon.

The "Nakaya" diagram in Fig. 3.3 displays the influence of temperature and humidity on the crystal shape. The vertical axis shows the density of water vapor in the excess of saturation with respect to ice. The curved line shows the saturation with respect to liquid water as a function of temperature. The borders between the different temperature ranges are very sharp — less than 1°C. Plates and prisms with planar surfaces are formed at low humidity. As the humidity increases, the edges and corners grow fastest and cavities are formed (the addition of water molecules from the environment and the disposal of the heat of crystallization is most effective at the edges and corners). As the humidity increases further, snow crystals with broad points are first formed, and these points gradually become narrower, and finally side branches are developed, resulting in so-called dendritic crystals at the highest humidity.

As a growing crystal is moved from one environment to another with different temperature or humidity, a mixed crystal will be formed, like those shown in Fig. 3.2. This is just what happens as a natural snow crystal encounters different atmospheric conditions when it falls to the earth. The resulting snow crystal is unique for its meteorological history; "the snow crystal is its own tachometer." Or, as Nakaya expressed it: "Snow crystals are the hieroglyphs sent from the sky."

At the low-temperature laboratory in Sapporo, snow research has continued under the leadership of Teisaku Kobayashi (Fig. 3.4) and present head Yoshinori Furukawa. A few examples of snow crystals grown in the laboratory are shown in Figs. 3.5a–l. The pictures may serve as an illustration of the gradual development of crystals with increasing humidity. The crystal with 12 branches appears to be a twin (*cf.* Descartes' pictures in Fig. 2.6).

Fig. 3.4. Teisaku Kobayashi at work growing snow crystals.

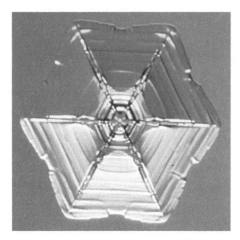

Fig. 3.5a

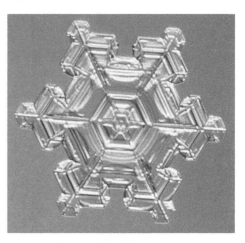

Fig. 3.5b

Fig. 3.5c

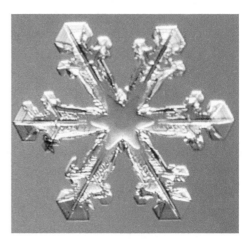

Fig. 3.5d

Fig. 3.5e

Fig. 3.5f

Fig. 3.5g

Fig. 3.5h

Fig. 3.5i

Fig. 3.5j

Fig. 3.5k

Fig. 3.5l

Fig. 3.6. Furukawa's snow hut for studying natural snow crystals.

Besides work in the low-temperature laboratory, Yoshinori Furukawa has followed up Nakaya's old field studies — taking pictures of natural snow crystals in a specially equipped "snow hut" (Fig. 3.6). Under the same conditions, the natural and artificial snow crystals are indistinguishable. Furukawa is in particular involved in experimental and theoretical studies of crystal growth and various surface phenomena. A comprehensive summary of his work appears in *Handbook of Crystal Growth* (Elsevier, 2015), Vol. I, pp. 1061–1112.

A neat way of studying snow crystals is illustrated below (Fig. 3.7). Don Komarechka has taken a large number of very beautiful macro photos in nature, some of which are reproduced below. A collection of his photographs is shown in the recent book *Sky Crystals: Unraveling the Mysteries of Snowflakes*. Exposure details and explanations of special features are also provided in the book.

Fig. 3.7. Study of snow crystals by Don Komarechka.

Scientific studies of snow and ice are today a very active field in many countries. Caltech physicist Kenneth Libbrecht has since childhood been fascinated by the beauty of snow crystals and has traveled the world studying and documenting them. He is also actively involved in taking photographs of natural snow crystals and in growing artificial crystals in the laboratory. He is the author of several books in the field, and a very large amount of information about the work by him and others on snow crystals is on the Internet: www.caltech.edu/~atomic/snowcrystals.

"Snow Crystals" with Seven and Eight Arms.   In Christmas decorations one often finds "snow" crystals with four, five, seven or even eight arms. People are of course free to choose any shape in the decoration — but if it is supposed to illustrate a snow crystal the only acceptable form is a pattern with six branches!

Can you find Two Identical Snow Crystals?   The answer to this classical question depends first of all on how closely you are able to look at the crystal. Using the naked eye or a microscope with optical resolution, it seems quite possible that you will find two crystals which you judge to be alike at this level of observation. However, at atomic resolution the answer will be quite different. It is easy to demonstrate that the probability must be extremely small that two crystals are built in exactly the same way. If we assume that a snow crystal weighs 10^{-4} g, such a crystal will consist of around 10^{18} water molecules. If an additional water molecule attaches, it will have around 10^{18} alternative positions to choose from (assuming a flat dendritic snow crystal and neglecting space problems). The probability that the next water molecule will attach to exactly the same place of an imagined identical snow crystal is clearly very small. Furthermore, the water molecules building up the ice structure may have different isotope compositions (with ^{16}O, ^{17}O, ^{18}O, ^{1}H or ^{2}D), which will lead to even more alternative compositions of the ice, and formally non-identical crystals.

Von Koch's Snowflake.   This is a purely mathematical construction but the shape has some similarity to natural snow flakes. It was one of the earliest fractal figures described and amongst the most important objects used by B. Mandelbrot for his pioneering work on fractals. It appeared in a 1904 paper by the Swedish mathematician Helge von Koch (Fig. 3.8), "*Sur une courbe continue sans tangente, obtenue par une construction géométrique élémentaire.*"

The von Koch snowflake is obtained by trisecting each side of an equilateral triangle and replacing the straight segment by

Fig. 3.8. Helge von Koch and the construction of his "snowflake."

two sides of a smaller equilateral triangle projecting outward, then treating the resulting figure the same way, and so on. The figure is self-similar, like all true fractals. The area is limited but the total length of the fully developed curve is infinite.

It seems that the term "fractal" is often used to describe richly branched objects in general, like dendritic structures in plants and trees. Should a snow crystal be characterized as fractal? If we assume that the tiny "sprouts" on the side branches of a fully developed dendritic snow crystal continue to grow in the same way as the big crystal and this procedure is repeated indefinitely, we could perhaps call it fractal. But you will never see such a fully developed snow crystal in practice, so dendritic snow crystals should probably not be characterized as fractal.

Ice Patterns on Windows.   When water vapor is deposited on a cold surface, such as a glass window, the feather-like patterns developed are quite different from natural snow crystals (Fig. 3.9). It is clear that scratches and impurities are to some extent responsible for the formation of these strange patterns. However, it is difficult to understand why an ice germ started at a certain point does not continue straight ahead from there. Free ice crystals grow normally in certain directions determined by the internal crystal structure. However, it appears that temperature and air humidity play an important role here. If the air is relatively dry and the surface very cold (down to $-25°C$), a pattern more closely resembling a snow crystal may develop. In air of higher humidity the surface is bombarded with such an amount of water molecules that a typical ice pattern does not have time to develop. Similar patterns may also form on a wooden deck. Here, the probability is larger that feather-like patterns will form on fat (hydrophobic) surfaces.

Twins, Snowflakes and Hail.   The snow crystals described so far are single crystals, i.e. the entire sample is continuous and there are no grain boundaries. A fully developed dendritic snow crystal is also a single crystal. However, if the

Fig. 3.9. Ice formation on cold surfaces.

growth of the crystal is disturbed, another crystal may develop with another orientation. In some cases such an outgrowth may have a direction which is related to the original structure and a twin or triplet is formed (see Fig. 3.5f).

As the snow crystal falls toward the earth, it will often hook onto other crystals in a random way and a *snowflake* is formed. Very large flakes are especially formed close to 0°C. A single snow crystal is transparent but a layer of snow looks white owing to repeated reflection of the light toward the randomly oriented snowflakes.

If a supercooled droplet of water larger than 0.01 mm comes into contact with an ice crystal, it may not have enough time to spread out over the crystal and result in a regular growth, and an irregular aggregate may form instead. On a large scale, when ice crystals are mixed with a thick cloud with supercooled water droplets, we get *hail*. Hailstones may become very large, and weights up to 1 kg have been reported. Compare that with the vision in Revelation **16**:21 — 50 kg!

Aging of Snow.   The snow on the ground gradually changes its appearance, depending on the temperature, pressure and mechanical treatment (for example when one is shoveling). The points of the dendritic snow crystals become more round and the snow more compact. Such changes occur already at temperatures far below 0°C. Small snow crystals are less stable than larger ones. In a mixture of crystals of different sizes, the large crystals will grow at the expense of the smaller ones. In glaciers this process may occur over several centuries, and extremely pure and perfect single crystals weighing several kilograms have been found.

Formation of Rain.   The formation of ice in the atmosphere is actually also a prerequisite for the formation of rain, according to present ideas. Most clouds consist of small droplets of water, and researchers have had difficulties in understanding how the droplets can unite into so large drops that they fall by their own weight. The individual water droplets have normally the same charge and repel one another. Gradually, it was realized that rain normally does not form until crystallization occurs: when some supercooled droplets freeze to ice, they continue to grow at the expense of other droplets. Finally, the large ice crystals begin to fall toward the earth, and if the temperature is sufficiently high they melt and *rain* falls. At our altitudes this is considered to be the most common mechanism for the formation of rain.

Snow Cannons.   *Snow cannon* can refer to two different things: (1) A device to make artificial snow in ski slopes (also called *snow gun*), or (2) The effect of heavy snow fall as cold air blows over a warm lake towards land.

(1) *Snow gun.*   Snow making started to be used on a commercial scale in the early 1970s. There are in principle two types of snow cannons. One type uses high pressure water and a powerful fan. The second type uses both water and air under

high pressure. A nozzle in the snow cannon forms droplets of water, so small that they remain suspended for a while so that they can grow and freeze. The ideal weather conditions for snow making are −5° to −15°C, less than 80% humidity and slight breeze. It is nowadays possible to make snow up to 0°C but the snow is then of slightly lower quality. A low humidity is important as a drop of water in dry environment partly evaporates and is then cooled down by the heat of vaporization. 10 cm of cannon snow corresponds to about 80 cm of natural snow. The freezing of the water is helped if you mix a nucleator of some kind into the water supply. The water may already contain some stuff that can act as nucleators, but many resorts also add special organic or inorganic materials as nucleating agents. The water is sometimes mixed with *ina* (ice nucleation-active) proteins from the bacterium Pseudomonas syringae. These proteins serve as effective nuclei to initiate the formation of ice crystals at relatively high temperatures.

Artificial snow is different from the natural snow in that it consists of small ice crystals and does not form flakes as natural snow. Artificial snow is also significantly harder in texture which can cause the adverse effect that it becomes easier "icy." Therefore this type of snow is also called *snice* (*snow-ice*). One advantage of the hard consistency is that it enables high speed skating, as it provides good grip for well-sharpened carving skis. Another advantage of the hard consistency is the durability; it melts much more slowly than natural snow which prolongs the season for ski resorts. As the properties are not the same as natural snow it may have implications for how one should wax the skies.

The strength of snice is almost like cement and snice is therefore used as building material, as e.g. when constructing ice hotels. The mixture used in the ice hotel in Jukkasjärvi in northern Sweden contains a larger amount of water than in normal artificial snow. The ice blocks are taken from the Torne river as this ice is especially clear (Fig. 3.10).

Fig. 3.10. From the ice hotel in Jukkasjärvi.

(2) *Lake effect snow.* When cold air blows over a warm lake toward land, a heavy snowfall can be a result. Such "snow cannons" (blizzards) are common in North America. When very cold air from Canada flows down over the Great Lakes, a huge amount of snow falls over especially the eastern shores of Lake Erie and Lake Ontario. The winter of 1880–1881 is considered the most severe ever known in the United States (see *The Long Winter* by Laura Ingalls Wilder). Figure 3.11 illustrates the situation on March 29, 1881 in Minnesota. In Sweden a notorious snowfall (although on a much lower scale) affected the Gävle area in early December 1998, when in three days 130 cm of snow fell, paralyzing all transport links in the area (Fig. 3.12).

Fig. 3.11. A train stuck in Minnesota in 1881.

Fig. 3.12. The snow cannon in Gävle in 1998.

4

Snow and Ice Crystals in Nature

Frozen water can appear in a seemingly infinite number of different shapes. A collection of the beautiful macro photos taken by Alexey Kljatov (Figs. 4.1a–f), Andrei Osokin (Figs. 4.2a–f) and Don Komarechka (Figs. 4.3a–l) is shown below.

Fig. 4.1a

Fig. 4.1b

Fig. 4.1c

Fig. 4.1d

Fig. 4.1e

Fig. 4.1f

Fig. 4.2a

Fig. 4.2b

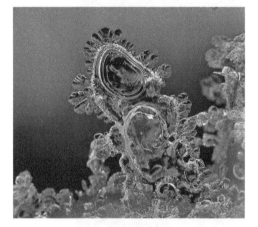

Fig. 4.2c

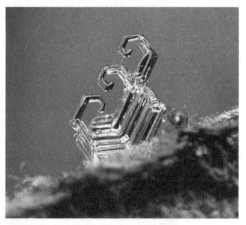

Fig. 4.2d

Fig. 4.2e

Fig. 4.2f

Fig. 4.3a

Fig. 4.3b

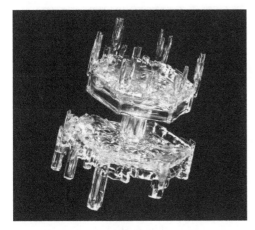

Fig. 4.3c

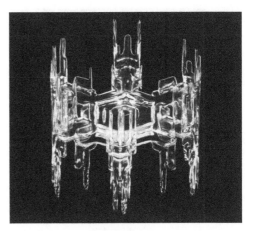

Fig. 4.3d

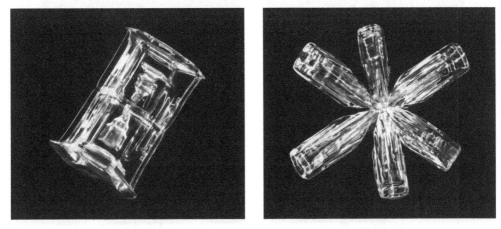

Fig. 4.3e

Fig. 4.3f

Fig. 4.3g

Fig. 4.3h

Fig. 4.3i

Fig. 4.3j

Fig. 4.3k

Fig. 4.3l

Thin Film Interference in Ice Crystals.   As an ice crystal grows, the edges and corners grow fastest as the addition of water molecules from the environment and the disposal of the heat of crystallization is most effective at these places. Small air pockets may then be left in the central parts of the crystal. In this way a very thin ice film may be formed between the outer surface and the air bubble. If light strikes the crystal, it is either transmitted or reflected at the outer surface. Light that is transmitted may again be reflected at the boundary between the thin ice layer and the air bubble (or at the bottom of the air bubble). The two reflected light beams will then interfere with each other. If the phase of the two reflected beams is the same, they will reinforce each other (constructive interference). If the phase is opposite, they will weaken each other (destructive interference). The phase difference depends on the thickness of the thin ice layer, the refractive index of the ice, and the angle of incidence of the light wave relative to the ice surface. Additionally, a phase shift of 180° will be introduced upon reflection at the boundary between air and ice. The pattern of reflected light which results from this interference will appear as colorful bands. If the blue light is quenched, yellow or reddish bands will appear.

Don Komarechka has taken a series of very nice macro photos showing very clearly this thin film interference (Figs. 4.4a–h). The pictures illustrate how the color changes with the thickness of the thin ice layer. A particularly interesting case is shown in Fig. 4.4g, where the crystal contains very small air bubbles distributed at different distances from the center of the crystal. As the angle of incidence relative to the air bubbles varies with the distance from the center, the color of the bubbles will also vary with this distance, as clearly shown in the picture.

Fig. 4.4a

Fig. 4.4b

Fig. 4.4c

Fig. 4.4d

Fig. 4.4e

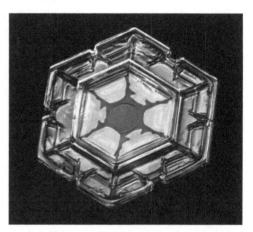

Fig. 4.4f

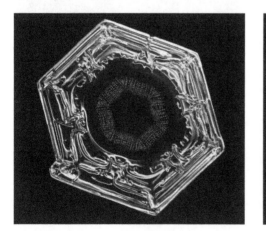

Fig. 4.4g

Fig. 4.4h

5
Snow for Pleasure and Art

Not only children enjoy making snow sculptures during the winter. Each year snow festivals are held all over the world. Two of the largest are in Sapporo in northern Japan and Harbin in northeastern China. Some examples from Sapporo are shown below. Even the Japanese army has in some cases been employed to build the largest monuments (Figs. 5.1a–f).

Fig. 5.1a

Fig. 5.1b

Fig. 5.1c

Fig. 5.1d

Fig. 5.1e

Fig. 5.1f

The temperature in Harbin stays below 0°C for several months during the winter (it sometimes drops to −40°C), and the fantastic constructions may then be enjoyed for a long time (Figs. 5.2a–e).

Fig. 5.2a

Fig. 5.2b

Fig. 5.2c

Fig. 5.2d

Fig. 5.2e

The British artist Simon Beck has created fantastic patterns in snowfields at Les Arcs in southeastern France and at Lake Louise in British Columbia. A very skilled orienteer, he forms the patterns by walking in snowshoes, using a sighting compass (as used by orienteering mapmakers) and a sketch of the pattern on a piece of paper. A typical pattern takes around 10 hours to produce. A selection of his pictures is reproduced (Figs. 5.3a–h). There is a book on sale specializing in his work; see http:// snowart.gallery.

Fig. 5.3a

Fig. 5.3b

Fig. 5.3c

Fig. 5.3d

Fig. 5.3e

Fig. 5.3f

Fig. 5.3g

Fig. 5.3h

A few beautiful frozen waterfalls in China are shown in Figs. 5.4a–d.

Fig. 5.4a. Blood Fairy Spring in Shijiazhuang, northern China.

Fig. 5.4b. Blood Fairy Spring in Shijiaxhuang, northern China.

Fig. 5.4c. Mancheng Longju Waterfall, central China.

Fig. 5.4d. Taihang mountains, northern China.

6

The Ice Surface and Formation of Ice Spikes

Ice Spikes.   A few years ago, when I was skating on a lake close to Uppsala, I noticed a lot of spikes sticking up everywhere (Fig. 6.1). Reports on the formation of ice spikes were already found in scientific journals in the early 20th century — among others, by O. Bally in 1935. Ice spikes are sometimes formed when a bowl of water is standing outside in cold weather. In recent years, a large number of photographs of natural ice spikes have appeared on the Internet, and a few years ago there was a lively discussion on how to explain the formation. Briefly, this is what happens.

The water first freezes around the edges on the top surface of the bowl, until a small hole is left unfrozen. At the same time, ice starts to form around the sides, inside the bowl. Since ice expands as it freezes (as the ice takes up a 9% larger volume), water is pushed up through the hole. The outgoing water freezes around the rim, forming an ice spike. The spike can continue to grow until all the water freezes. As the ice spike is formed when water is pushed up through a hole in the growing ice, the limited space in the bowl helps to build up the pressure needed. Accordingly, ice spikes are most easily formed in small containers. Small ice spikes

Fig. 6.1. Ice spikes on a lake in the neighborhood of Uppsala.

can occasionally also be found on ice cubes in refrigerators, using distilled water in plastic ice cube trays. The ice spikes are mostly thin and cylindrical, but triangular and inverted pyramids have also been reported. The angle between the spike and the surface varies, and does not seem to have any direct relation to the internal crystal structure.

It is somewhat surprising that so many ice spikes were formed all over the lake as shown in Fig. 6.1. A large number of small pinholes evidently remained as ice formed on the surface and water pressure was somehow built up below (probably owing to the thickening of the ice layer).

The spikes normally become only around 5–6 cm long but occasionally much longer spikes have been reported. The following article, "Due North," by Harold L. Kirk, appeared in *Harborcreek Historical Society Newsletter*, April/May 2007, p. 6 (reprinted with permission):

> At 8 a.m. on a cold Saturday morning in March of 1963, Gene Heuser left his warm home on East Lake Road and headed due north. When he reached the shore of Lake Erie, the only thing ahead of him was an icy barren wasteland as far as the eye could see. His plan was to hike over the ice, about 32 miles, to the Canadian lighthouse at Long Point, stay the night and then hike back to Harborcreek. Little did he envision the many obstacles that lay ahead. Using only a small compass to guide him, he soon found that heading due north was seldom possible as he encountered pillars of ice five feet high and snow drifts of over 10 feet. He told a reporter later, "I never expected to see what I saw. It was not just a smooth surface."
>
> As evening approached and he was still miles from land, he knew he would be spending a long cold night on the ice! The moon shone brightly for about an hour but later, clouds covered the sky leaving him in near total darkness. Using a small flashlight allowed him to continue his northward trek. He said, "The flashlight lit up these huge ice chunks with a fluorescent glow into eerie forms and shapes like those of a barren planet. Sometimes I fell on the jagged surface and just lay there on the ice. I knew I could not lie down long or I would freeze." He also said that one of the most vivid recollections of that long night was of the small pinholes in the ice through which the water below was periodically forced under pressure to spout up into the air and freeze. The frozen spurts looked to him like telephone poles standing straight up all over the lake. He told the *Erie Morning News* later, "I knew my planned route from Shade's Beach to Long Point was about 32 miles but I figured I must have walked over 50 miles because of the drifts and ice chunks I had to walk around."
>
> Well past daybreak on Sunday morning, Gene reached the lighthouse where he saw some shacks belonging to a team of Canadian scientists making a lake study. They didn't believe that he had just strolled over from the nation to the south until he showed them his identification. Canadian police escorted him to the mainland at Port Rowan, Ontario. Gene quickly revised his original plan of walking back to Harborcreek and instead called his brother in Buffalo to pick him up.

The Ice Surface.   Notice that the ice surface in Fig. 6.1 is quite uneven. This is a common feature of almost all ice surfaces.

When water in a closed environment, such as a small container, begins to freeze, separate ice crystals will form at the edges and the growing ice crystals will soon collide with one another. At the same time the water below continues to freeze and pressure is built up from below and the ice on the surface cannot form a completely flat surface, as illustrated in Fig. 6.2 (this may be considered the first stage of spike formation, in which case water is pressed through small pinholes still remaining in the surface). For this reason, the ice surface on a small pond in the street or even on an inland lake will not be completely flat throughout. The best condition for finding an extended completely flat ice surface is consequently at the seashore, where the ice may grow outward without colliding with other growing crystals.

Fig. 6.2. Typical pattern of the ice surface formed in a container.

Icebergs.   Icebergs form when chunks of ice "calve" or break off from glaciers. They come in all shapes and sizes, from small chunks to the size of a small country. Only 20% of a floating iceberg is visible above water, owing to the difference in density of ice and water. Most icebergs are found in the North Atlantic and the cold waters surrounding Antarctica, and pose a danger to ships — the sinking of the *Titanic* in 1912 is still fresh in the memory. But there have also been plans to utilize them for practical purposes. The possibility of towing icebergs to arid regions with a lack of fresh water has been considered for several decades, and such transportation has been tried in a few cases. For example, between 1890 and 1900, small icebergs were towed by ship from Laguna San Rafael, Chile, to Valparaiso and even to Peru, a distance of 3,900 km. Although such transportation is technically possible, the economy is doubtful. One major problem is that a large part of the iceberg melts during the transportation.

Effect of Ice Melting on the Sea Level.   With the present increase in the global temperature, which may lead to rapid melting of ice in Arctic regions, there has been great concern that it will result in a large increase of the sea level. Here one should perhaps be reminded that only the melting of land-based ice will affect the sea level. The melting of sea ice and floating icebergs will not change the sea level.

Ice as Aircraft Carrier and Project Habakkuk.   During the Second World War, a top secret project was launched by the British to build a ship from ice to be used as a landing strip for aircraft and to be used against German U-boats in the mid-Atlantic, which was beyond the flight range of land-based planes at that time. The idea came from Geoffrey Pyke, who worked for Combined Operations Headquarters. Pure ice was evidently too brittle and too quickly melting for the purpose. However, the physical chemist Herman Mark found that a mixture of ice and wood pulp makes a very hard material, which was named pykrete. It could be machined like wood and when immersed in water formed an insulating shell of wet wood pulp on its surface that protected its interior from further melting. Churchill approved the project named Habakkuk. The name originates from one of the books in the Hebrew bible (Habakkuk 1:5: ... "be utterly amazed, for I am going to do something in your days that you would not believe, even if you were told"). Two crystallographers, John Bernal and Max Perutz were involved as scientific advisors in the project.

Pyke was not the first to suggest a floating mid-ocean aircraft carrier made of ice. A German scientist, Gerke von Waldenburg, had proposed the idea and carried out some preliminary experiments on Lake Zurich in 1930. In 1940, the idea for an ice island was circulated around the British Admiralty, but was treated as a joke by officers, including Nevil Shute, who circulated a memorandum that gathered very sarcastic comments.

In early 1942, Perutz was asked to determine whether an ice floe large enough to withstand Atlantic conditions could be built fast enough. Perutz pointed out that natural icebergs have too small a surface above water for an airstrip, and are inclined to suddenly rolling over (*cf.* the chapter on icebergs). The project would have been abandoned if it had not been for the invention of pykrete. However, Perutz found a problem: ice flows slowly (plastic flow), and his tests showed that a pykrete ship would slowly sag unless it was cooled to $-16°C$. To accomplish this, the ship's surface would have to be protected by insulation, and it would need a refrigeration plant and a complicated system of ducts.

The decision was made to build a large-scale model at Jasper National Park in Canada to examine insulation and refrigeration techniques, and to see how pykrete would stand up to artillery and explosives. Large ice blocks were constructed at Lake Louise, Alberta, and a small prototype was constructed at Patricia Lake, Alberta, measuring only 18 meters by 9 meters, weighing 1,000 tons and kept frozen by a one-horsepower motor. The workers were never told what they were building. Bernal informed Combined Operations Headquarters that the Canadians were building a model, and that it was expected to take eight men

Fig. 6.3. Proposed aircraft carrier made out of pykrete.

14 days to build it. Churchill asked the Chiefs of Staff Committee to arrange for one complete ship at once, with the highest priority, and that further ships were to be ordered immediately if it appeared that the scheme was certain of success. The proposed aircraft carrier in action is shown in Fig. 6.3.

The Canadians were confident about constructing a vessel for 1944. The necessary materials were available to them in the form of 300,000 tons of wood pulp, 25,000 tons of fiberboard insulation, 35,000 tons of timber and 10,000 tons of steel. The cost was estimated at £700,000.

By May, the problem of sagging of the vessel (cold flow) had become serious and it was obvious that more steel reinforcement would be needed, as well as a more effective insulating around the vessel's hull. The Canadians decided that it was impractical to attempt the project "this coming season." Bernal and Pyke were forced to conclude that no Habbakuk vessel would be ready in 1944.

Shooting Incident. According to some accounts, at the Quebec Conference in 1943, Lord Mountbatten brought a block of pykrete along to demonstrate its potential to the admirals and generals who accompanied Winston Churchill and Franklin D. Roosevelt. Mountbatten entered the project meeting with two blocks and placed them on the ground. One was a normal ice block and the other was pykrete. He then drew his pistol and shot at the first block. It shattered and splintered. Next he fired at the pykrete to give an idea of the resistance of that kind of ice to projectiles. The bullet ricocheted off the block, grazing the trouser leg of Admiral Ernest King, and ended up in the wall.

Perutz gave an account of a similar incident in his book *I Wish I Made You Angry Earlier.* A demonstration of pykrete was given at Combined Operations Headquarters by a naval officer, Douglas Grant, who was provided by Perutz with rods of ice and pykrete packed with dry ice in thermos flasks and large blocks of ice and pykrete. Grant demonstrated the comparative strength of ice and pykrete by firing bullets into both blocks: the ice shattered, but the bullet rebounded from the pykrete and hit the Chief of the Imperial Staff (Sir Alan Brooke) in the shoulder; he was luckily unhurt.

End of Project.   Later in 1943, Habbakuk began to lose priority. Mountbatten listed several reasons.

- Demand for steel for other purposes was too great.
- Permission had been received from Portugal to use airfields in the Azores, which facilitated the hunting of U-boats in the Atlantic.
- The introduction of long-range fuel tanks allowed British-based aircraft extra patrol time over the Atlantic.
- The numbers of escort carriers were being increased.

The final meeting of the Habbakuk board took place in December 1943. It was officially concluded that "The large Habbakuk II made of pykrete has been found to be impractical because of the enormous production resources required and technical difficulties involved."

It took three hot summers to completely melt the prototype constructed in Canada!

7

Structure and Physical and Chemical Properties of Water and Ice

The Water Molecule. As oxygen is more electronegative than hydrogen, the electrons are pulled slightly toward oxygen, and hydrogen gets a small net positive charge. The oxygen gets a corresponding negative charge (Fig. 7.1). This results in polar properties of the water molecule (it has a dipole moment of 1.85 debye). Many of the special features of the water molecule are due to this polarity. Of particular importance is the distinctive tendency of the water molecules to form hydrogen bonds with one another or with other polar molecules, and they can then operate both as donors and acceptors. Sometimes the two lone pairs in the water molecule are illustrated as "rabbit ears" standing out tetrahedrally relative to the O–H bond directions (Fig. 7.2).

However, the actual electron distribution does not at all agree with this model. There is a fairly even distribution over the entire area. The electron distribution of the free water molecule from quantum-mechanical calculations is shown in Fig. 7.3: (a) in the plane; (b) perpendicular to the plane of the molecule. The corresponding deformation density is shown in Fig. 7.4 (the deformation density shows the deformation of the electron clouds of the free oxygen and hydrogen atoms when the molecule is formed — the difference between the actual density minus the "promolecule" density).

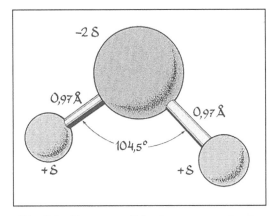

Fig. 7.1. Geometry of the free water molecule.

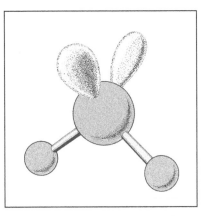

Fig. 7.2. "Rabbit ear" model.

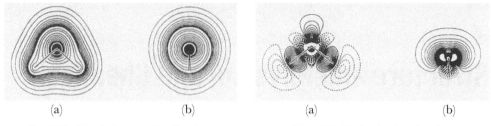

(a)   (b)   (a)   (b)

Fig. 7.3. Total electron distribution.   **Fig. 7.4.** Deformation density.

As the simple picture of the water molecule with "rabbit ears" does not illustrate the actual electron distribution, this model should be avoided. With the rabbit ear model one might suggest that the water molecule binds specifically in the free electron pair directions (tetrahedrally relative to the OH directions) owing to an electron concentration in these directions. It is true that the water molecule often forms a tetrahedral arrangement with other molecules, but this can be explained by purely topological reasons: Suppose that we want to build a three-dimensional arrangement of water molecules and that we require that all these have the same environment. If each water molecule functions as donor (O–H···) in two hydrogen bonds (O–H···O), then each water molecule also must accept two hydrogen bonds. For purely geometrical reasons, these two acceptors will be arranged approximately tetrahedrally relative to the OH directions in order to form the most favorable arrangement.

For a more exhaustive discussion on the role of the lone pairs in hydrogen bonds in general, see Chapter 17.

The Structure of Ordinary, Hexagonal Ice, I_h. In all forms of ice, each water molecule is tetrahedrally surrounded by four other water molecules (Fig. 7.5). Owing to the hydrogen bonds, the O–H bond is slightly longer in ice than in the free water molecule. The only structure that is stable at ordinary pressure and at moderately low temperature is hexagonal ice, I_h, which has sixfold internal symmetry. The crystal structure is shown in Fig. 7.6. There is one very important feature of this

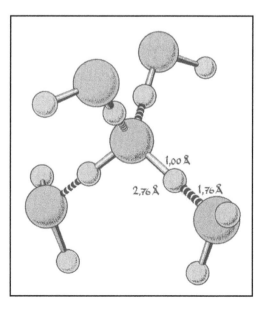

Fig. 7.5. Environment of H_2O in ice.

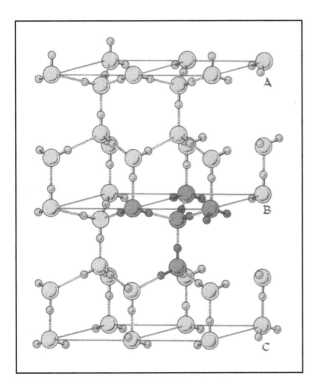

Fig. 7.6. Structure of ordinary ice.

structure: the water molecules are disordered! The picture shows an example of how the molecules can be oriented differently in different unit cells. Compare the molecules marked A, B and C. If all possible orientations are equally frequent in the structure, an arrangement O−H···O and O···H−O will be equally probable in a particular hydrogen bond (if all unit cells are taken together). A diffraction study will then show one-half hydrogen atom at each place. This disorder is maintained all the way down to 0 K, and this results in a so-called residual entropy (see more details below).

Ordinary ice has in many ways unusual properties compared to similar substances containing light molecules. As ice has so many interesting physical properties, it was natural that many scientists tried to determine its crystal structure very early on. But that proved more difficult than expected. Already in the early 1920s, Bragg and other crystallographers showed that ordinary ice has a hexagonal symmetry (D6h, P63/mmc), and the positions of the oxygen atoms were determined. The structure forms a very open network, due to the hydrogen bonds. If the molecules are instead tightly packed next to one another, each molecule will be surrounded by no less than 12 neighbors. Such a packed arrangement occurs in the crystals of the analogous compounds H_2S, H_2Se and H_2Te. In these cases no strong hydrogen bonds are formed and the molecules try to pack as closely as possible.

There was general agreement about the positions of the oxygen atoms, but the placement of the hydrogen atoms caused a lively debate. The problem was that at that time only X-ray diffraction was available and as the hydrogen atoms spread X-rays rather weakly they could not find the hydrogen atoms. Today the X-ray method is much more developed and the location of hydrogen atoms can be determined in not-too-complicated compounds, though still with moderate accuracy. Many more-or-less fanciful models were proposed. Some researchers suggested that the water molecules do not have the same orientation in every unit cell, which is not possible in an ideal crystal structure. Thermodynamic measurements showed that when ice is cooled to near 0 K there remains a certain entropy, 3:41 J/degree·mol, a so-called residual entropy. Entropy is a measure of the degree of disorder in a system: in a completely ordered system, the entropy must approach zero at 0 K. The experimental measurements thus indicated a disordered structure. Pauling suggested that water molecules could have a statistically equal distribution among all possible orientations of the water molecules, while still forming hydrogen bonds to the four neighbors (see Fig. 7.6). Based on this model he calculated an entropy of 3:38 J/degree·mol, which is close to the experimental value. The dispute about the correct ice structure was long and lively. The definitive breakthrough did not come until we had access to neutron diffraction.

At the end of the 1940s, one of the first research reactors in the world was built in Oak Ridge, USA, and here pioneering work was done in neutron diffraction research. E. O. Wollan, W. L. Davidson and C. G. Shull made in 1949 a study of ice in powder form. In 1994 Shull received together with Bertram Brockhouse the Nobel Prize in Physics for this neutron study of ice, among other work. Even more detailed information was obtained some years later when S. W. Peterson and H. A. Levy in Oak Ridge studied single crystals of ice. In both of these early studies heavy water, D_2O, was used as the deuterium isotope gives only a small contribution of disturbing (incoherent) scattering compared to ordinary hydrogen. Peterson and Levy are probably the first to investigate a single crystal by neutron diffraction.

The neutron results were awaited with great excitement — the question of the ice structure had now become a classic problem. The investigations confirmed Pauling's theory of the disordered structure of ice. A diffraction study does not show the content in a specific unit cell, but the mean value of the contents in all unit cells. In a completely disordered structure the probability is only 50% that hydrogen is in one of the two possible positions in a particular oxygen–oxygen bond. In the neutron study it was indeed found that the hydrogen positions were on average only half-occupied. Nevertheless, there are still many questions about the details of ice structure to explore.

Thermodynamic measurements of $Na_2SO4 \cdot 10H_2O$ have shown that this compound also retains a certain residual entropy when cooled to low temperatures. During my time in Berkeley, 1957–1959, the crystal structure of this compound was determined by X-ray diffraction. In

Fig. 7.7. Alternate configurations for water molecules.

the structure there are two kinds of rings, each with four hydrogen bonds. There are two possible configurations of the hydrogen atoms in these rings (Fig. 7.7) and complete disorder of these configurations corresponds to the residual entropy found experimentally ($R \ln 2$ per mole).

The Structure of Other Forms of Ice. Depending on temperature and pressure, water can assume about 17 different crystal structures — perhaps more than any other known material. The phase diagram is shown in Fig. 7.8.

Amorphous Ice and Ice I_c. If water vapor condenses on a cold surface at temperatures below about −160°C, a non-crystalline, amorphous form — glass — is created. If the condensation occurs between about −160°C and −120°C, a crystalline, cubic form — ice I_c, with a diamond structure is created, but otherwise it is similar to ice I_h. Cubic ice possibly also occurs at high altitude in the atmosphere.

Ice II–XI. These structures are formed at higher pressures — between 0.2 GPa and 2.2 GPa — except for ice X, which is first formed at pressures above 44 GPa. The water molecules in ice II, VIII, IX and XI are more-or-less ordered, whereas the other crystal structures are partially or completely disordered. The coordination is in all cases tetrahedral, as in hexagonal ice, but the relative positions of the tetrahedra are different and the water molecules are more densely packed in the high-pressure forms. The structure of ice X is not yet known but is postulated to be of an entirely different character with symmetrical hydrogen bonds.

Doping ice I_h with 0.001–0.1 moles of KOH has produced crystals with an orderly proton distribution corresponding to the structure of ice XI.

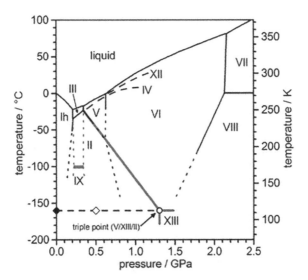

Fig. 7.8. Phase diagram of ice.

8

Physical Properties of Water and Ice; Significance in Nature

The melting and boiling points of water are unusually high owing to the hydrogen bonds. Without hydrogen bonds the melting point should be around $-100°C$ and the boiling point perhaps around $-80°C$, when compared with the analogous compounds H_2S, H_2Se and H_2Te, where the bonds between the molecules are much weaker (van der Waals bonds), and if the molecular weight of the water molecule is taken into consideration. To melt 1 kg of ice, 334 kJ is needed (80 cal/g), and the corresponding amount of heat is released when water freezes. The melting and boiling points of simple hydrides are compared in Figs. 8.1 and 8.2.

An interesting phenomenon occurs in the fall when the water in a lake starts to freeze *and there is no wind at all*. When some water turns into ice, the heat released is transferred to other water molecules, which may then turn into gas form. As the air is cold, this gas will condense to water droplets and mist is formed — "the lake smokes." According to the old tradition, you can then tell that the lake has started to be covered with ice. It should perhaps be added that "smoke" may of course also be formed when cold wind blows over a lake with warm water — and perhaps cause a snow cannon.

The water is very easily supercooled: when ice is formed the irregular network of hydrogen bonds in liquid water must be reorganized into a regular tetrahedral network. Clean water can thus be supercooled down to $-20°C$ before it crystallizes, and water droplets (1–10 microns) in the atmosphere even down to $-40°C$.

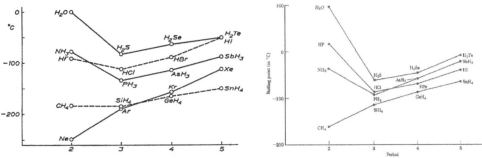

Fig. 8.1. Melting points of simple hydrates.   **Fig. 8.2.** Boiling points of simple hydrates.

The very special characteristics of the ice structure make it difficult to take up other compounds; an exception is ammonium fluoride, whose structure is similar to that of ice. When ice is formed from saline water, the majority of the salts will accordingly remain in the solution and the ice will become virtually salt-free. This can in principle be used for the production of drinking water from seawater, but this method of desalination is very energy consuming. However, if the ice formation takes place very quickly, the salts will accompany the ice crystals as an impurity, although not as a solid solution inside the ice structure. The principal current process to separate salts from water uses semipermeable membranes and pressure, applying reverse osmosis technology.

The very open structure of ice leads to a low density — 0.917 g/cm^3 — of pure ice at 0°C. When ice melts at 0°C, part of the bonds in the tetrahedral network are broken and the water molecules can pack somewhat closer (but hydrogen bonds still play a major role in liquid water, and much of the short-range order persists even when the ice has melted). The density of liquid water at 0°C — 1.000 g/cm^3 — is therefore higher than for ice — a remarkable fact, as in most cases a solid has a greater density than the corresponding liquid. Ice will accordingly float on water, which is of great importance in nature, as we shall see. When water freezes to ice, the volume increases by about 9%, and in closed systems this will have strong expansion effects — "frost shattering" in nature. For the same reason, the melting point of ice is lowered when subjected to pressure (0.0074°C/bar), which is important in glaciers, for example.

When the temperature of water rises from 0°C, additional hydrogen bonds between the water molecules are broken and therefore the density increases at first. At the same time, however, another factor comes into play: when the temperature rises, the motion of the water molecules increases and the molecules will take up more space (the vast majority of solids and liquids expand with temperature for the same reason). Eventually, the latter effect will dominate, and water expands with temperature in the same manner as a normal liquid. The combined effect of these two factors has the result that pure water gets a maximum density at +4°C — a most remarkable phenomenon that is of great importance in nature. With increasing salinity the freezing point decreases as well as the temperature of the maximum density, so that water with a salinity of 2.5% on cooling constantly increases its density down to the freezing point (for the oceans with a salinity of 3.5%, the freezing point is −2°C).

When the water in a lake gradually cools down in the fall, the following events will occur owing to the special properties of ice and water (if we disregard the effects due to wind). When the surface water is cooled, it will sink toward the bottom, until the bottom water has a temperature of +4°C. Upon further cooling from +4°C to 0°C, the water at the surface will remain there as it has a lower density than the bottom water. Eventually, ice will be formed, and since it does not sink to the bottom the ice will form an insulating layer that delays the continued cooling of the water. The growth of the ice layer occurs all the time on the underside of the ice. Ground freezing is prevented effectively and normally occurs only in shallow water (the effects of currents can of course accelerate this process). It is quite remarkable how this collaboration of several unusual physical properties of ice and water affects our environment and helps aquatic animals to survive the winter.

While the serum of fish living in polar seawater can carry enough salt to lower their freezing temperature by about 1°C, this is not enough to prevent freezing. Consequently, they must rely on another mechanism for survival in a supercooled state. The antifreeze effect is a crucial issue and is related to crystal growth controlled by biological macromolecules.

Surface Properties of Ice and Snow. Why do we glide more easily on ice and snow than on asphalt or gravel? A common notion has been that when the skate glides across the ice, the pressure causes melting of the ice and provides a liquid film with very low friction. But this explanation is not entirely correct, according to current research. When we go skiing or skating, the pressure is too low to melt the ice. Another theory been suggested regarding the structure of the outermost layer: the water molecules on the surface do not have a complete system of hydrogen bonds and the incomplete bonding situation could perhaps allow the molecules on the surface to rotate. Individual molecules or small groups of molecules could then function as a sort of ball bearings. However, it seems more likely that the molecules turn aside but are still bound to the underlying molecules.

Research on the surface properties of ice remains very active. Extensive experimental investigations have been made in recent years, e.g. by ellipsometry, X-ray scattering, LEED (low energy electron diffraction), NMR (nuclear magnetic resonance) and AFM (atomic force microscopy). Theoretical calculations — among others, the application of molecular dynamics simulation — has also given detailed information about the arrangement and dynamics of water molecules in ice and water. These results show that the water molecules of the outermost layer have great mobility and increasing disorder with increasing temperature. It is thus a question of a more-or-less disordered network of water molecules of high mobility at the surface but not a typical liquid layer where the molecules are relatively free to move; we can call it a "quasi-liquid layer." The surface properties of most materials can be quite different from the inner part of the material, "bulk properties," and approach the properties of the liquid phase. The ice surface may perhaps be likened to a brush: the hairs are stiff at low temperature and this leads to a rough surface, but the hairs bend more easily when the temperature rises and the friction is then lower when an object such as a ski glides over the "brush."

The Structure of Liquid Water. How are the water molecules arranged in water? They move around or twist in the liquid, on a timescale as short as a picosecond (10^{-12} s), and since the arrangement can change more or less radically, it is obviously impossible to report a specific, fixed "structure." To determine the instantaneous structure we must employ experimental methods which can register the structure in an even shorter time. X-ray and neutron diffraction (elastic scattering) only give information about the average distribution of molecules over a long time span (a superposition of all structures during this time), but on the other hand they can give definite information about the distances between the molecules (the radial distribution). Since diffraction and spectroscopy have perhaps provided the most important contributions to the models of the water structure, we will touch upon the information these methods can provide.

Radial Distribution. The electron distribution in atoms is usually illustrated with the so-called radial distribution function, $4\pi r^2\psi^2$, which indicates the relative probability of finding the electron at a distance r from the center. Similarly, the distribution of molecules in a liquid is described by a radial distribution function, $g(r) = 4\pi r^2\rho(s)$, which indicates the relative probability of finding two atoms at a distance r from each other (the "pair correlation"). Maxima occur where the probability is very high. However, the diffraction investigations cannot give direct information on which pair of atoms is at this particular distance.

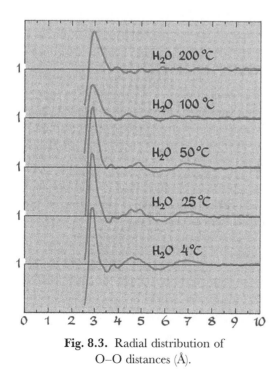

Fig. 8.3. Radial distribution of O–O distances (Å).

X-ray Diffraction. X-rays are scattered by electrons, and therefore heavier atoms scatter the radiation more strongly than light atoms. The contribution to $g(r)$ is proportional to the product of the scattering power of the two atoms in the atom pair. In the case of water, oxygen scatters the X-rays about eight times more strongly than hydrogen with only one electron. This means that the contribution to $g(r)$ is completely dominated by the oxygen–oxygen pair. The relative probability of finding neigboring water molecules at a certain oxygen–oxygen distance at different temperatures is shown in Fig. 8.3. The large peak at 2.8 Å corresponds to the O–O distance between the closest hydrogen-bonded water molecules. From the location of the smaller peaks at larger distances, one tries to draw a conclusion about the arrangement of water molecules further out and to derive a model for the water structure.

The distance of 2.84 Å in liquid water is slightly longer than the distance in ice, 2.76 Å. These values agree well if we assume that the regular tetrahedral network in ice is partly broken down in liquid water so that each water molecule on the average is surrounded by slightly more neighbors than in ice (approximately 4.5 compared to 4 in ice, and each distance then becomes somewhat longer).

Neutron Diffraction. Neutrons are scattered by the nuclei, and light nuclei may scatter about as strongly as heavier nuclei. This means that the hydrogen–hydrogen contribution to $g(r)$ will be approximately as large as the oxygen–oxygen contribution. The radial distribution function will thus reproduce all distances, O–O, O–H and H–H, with approximately equal weight. However, with all these three types of distances simultaneously included, the $g(r)$ plot becomes extremely complicated and practically impossible to interpret. But, through replacement of normal hydrogen (H) by deuterium (D), one can vary the relative contributions of $g(r)$. The scattering power of hydrogen and deuterium is very different, and by diffraction studies of water with three or more different isotopic compositions it is then possible to separate $g(r)$ for the three different types of distances. In recent years, neutrons have mainly been utilized in the diffraction investigations.

Models for the Water Structure.

Many different models have been suggested for how water molecules are arranged in liquid water, and there is still a lively debate. Many of the earlier models only tried to explain specific anomalous properties of water, such as the high heat capacity and compressibility, or the increase in density when the ice melts. But it has been difficult to find a model that can explain all the experimental facts in a reasonably satisfactory manner. It is particularly important to find a model that is consistent with the current "structure-specific" data, where the results from diffraction and spectroscopy are vital. Essentially two main types of models were suggested early on: "homogeneous models" (or "uniform continuum models") and "cluster models" (or "mixture models"). These models are briefly described below.

Homogeneous Models. The original model was formulated by J. D. Bernal and R. H. Fowler in a classic work published in 1933 and by J. Lennard-Jones and J. A. Pople in 1951. Here it is assumed that all water molecules are all the time hydrogen-bonded to one another in a network without regular repetition, and where the molecules retain a fourfold coordination even when the details of the arrangement change with time by the bending and stretching of the hydrogen bonds. It is thus assumed that none of the four hydrogen bonds around a certain water molecule is broken when the molecules move, and that the bending and stretching of the bonds is conducted independently in the different water molecules.

Cluster Models. Here it is suggested that clusters of three- or four-coordinated water molecules are bound together in a rigid network with a lifetime of about 1 nanosecond (10^{-9} s). It is further assumed that these clusters are separated by borders in which the water molecules are involved in only one hydrogen bond or perhaps even none at all. In the "flickering cluster" model by H. S. Franks, the boundaries and the arrangement of the clusters change over time by cooperative movements of the water molecules. In contrast to the homogeneous models, in cluster models it is assumed that hydrogen bonds are not so easily bent and that linear hydrogen bonds have a significantly lower energy than bent ones.

An alternative model, which can best be attributed to the cluster models, is the "interstitial model." Here it is assumed that the individual water molecules can be shaken loose and enter cavities in the very open ice structure. Such a model was proposed, among others, to explain the increase in density when the ice melts.

Theoretical Calculations. With the advent of fast computers, it has become possible to make advanced theoretical calculations of the water structure. In so-called molecular dynamics simulations, a starting structure is first selected (such as the arrangement of ordinary ice) and subsequently the individual water molecules are allowed to move in accordance with Newton's equations of motion, and thus in a purely classical physical manner. However, the problem is to find a good model for the forces between the water molecules. In the calculations it is important, among other things, to take into account "many-body interactions" ("cooperativity"): when two water molecules are bonded to each other, the electron distribution in these molecules is affected and this alters the forces to other water molecules. The theoretically calculated radial distribution from molecular dynamics simulation is in good qualitative agreement with the experimentally determined distribution.

Present Situation. Neither of the models described above seems to be separately sufficient to explain all the anomalous properties of liquid water. Against a strictly uniform model speaks, among other things, the fact that the water molecules most likely have an environment that varies strongly with time owing to rotation of the water molecules and breaking of the hydrogen bonds. Against a rigid cluster model with close-to-linear hydrogen bonds speak surveys of crystalline hydrates which show that the hydrogen bonds in most cases are more or less bent. In the cluster models it has to be assumed that the clusters are constantly changing rapidly and the "unbound" water molecules between the clusters can reasonably exist only for a very short time, of the order of one-tenth of a picosecond, for example in connection with reorientation. On a slightly longer timescale, it is very unlikely that a water molecule is not hydrogen-bonded to any neighbor at all; from structure studies of compounds containing water, it is known that water molecules have a *very strong* tendency to be involved in hydrogen bonds. From molecular dynamics simulations and experimental studies, we also know that the water molecules have on average an environment reminiscent of the tetrahedral arrangement of ice and where the network is still relatively open.

Water, a Solvent with Many Interesting Properties

Ambient Water. Water is called the "universal solvent" as it is capable of dissolving more substances than any other liquid. Most people probably think of water as a completely non-reactive and innocent, non-toxic compound. After all we may drink water without any harmful effects (in limited amounts). But water is actually surprisingly "corrosive" to matter in general and attacks a large number of materials. These properties are intrinsically due to the polarity of the water molecule: The lone pair region may attract positively charged ions and molecules, and the hydroxyl group may bind to negatively charged molecules and ions. Water may accordingly interact with a great number of compounds in different ways. On the other hand, water is a poor solvent for non-polar organic substances.

A great number of more or less drastic events in nature are caused by water. From geology, we know that after some million years high mountains may be broken down by the action of water. The freezing and melting processes of ice are then also responsible in this context. Transformations of our earth by water and ice are treated in Chapter 18.

Supercritical Water. Above the critical point, a temperature of 374°C and a critical pressure of 22 MPa, the liquid and gaseous phases of water become identical. The properties are dramatically different in the supercritical state: The fraction of hydrogen-bonded molecules is greatly decreased, as well as the viscosity, dielectricity, etc. All such drastic changes have as a result that *supercritical water behaves much less like a polar solvent*: Non-polar organic molecules become soluble whereas salts precipitate out of the solution! The solubility of various organic compounds makes it possible to produce a variety of functional materials without using polluting organic solvents. Supercritical water has replaced hexane in many solvent extraction processes. Hazardous waste may be destructed by oxidation in supercritical water in the presence of oxygen, etc. Supercritical water may become the cleanest solvent of all, with myriad applications!

In this context should also be mentioned the large industrial importance of supercritical carbon dioxide. The relatively low temperature needed in these processes and the stability of CO_2 allows many compounds to be extracted with little damage.

Why is Water Blue?   The observed blue color of a lake is caused both by the reflected skylight and the intrinsic color of the water. The contribution by the reflected skylight depends on the observation angle. At a distance, the color of the reflected skylight dominates; looking more directly down, the intrinsic water color dominates. In the following we will only discuss the intrinsic color of water.

Pure water in a cup appears colorless; in a swimming pool it looks light blue. With increasing depth of the water in the pool, the blue color becomes stronger and stronger. Most substances owe their colors to the interactions of visible light with the electrons. However, for water all its electronic absorptions occur in the ultraviolet region. In the visible spectrum, the absorption originates from vibrational transitions (involving high overtone and combination states of the vibrational modes). To my knowledge the intrinsic blueness of water is the only example from nature in which color originates from vibrational transitions. In contrast, D_2O is colorless because all of its corresponding vibrational transitions are shifted to lower energy by the increase in isotope mass.

The visible part of the spectrum extends from 390 nm to 700 nm, with a minimum in the absorption at 420 nm, corresponding to blue light (Fig. 8.4). *From the blue to the red part of the spectrum there is a steady increase in absorption and this explains the intrinsic blueness of water.*

(For a quantitative calculation of the absorption: The absorption coefficient a at a particular wavelength is given by the equation $I/I_0 = e^{-aL}$, where I is the transmitted intensity of the light, I_0 is the incident intensity and L is the path length (in cm). For blue light $a \sim 10^{-4}$ cm^{-1} and for red light $a \sim 10^{-2}$ cm^{-1}.)

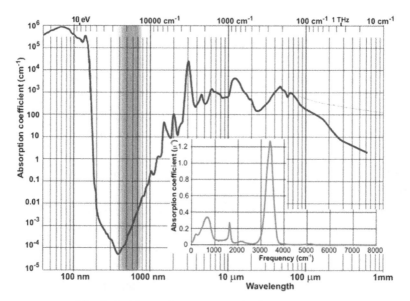

Fig. 8.4. Absorption spectrum for liquid water.
(© Martin Chaplin, Water Structure and Science. Permission granted.)

The penetration depth in water depends on the wavelength. The long wavelengths of the visible spectrum — red, yellow, and orange — can penetrate down to approximately 15, 30, and 50 meters, respectively, while the short wavelengths — violet, blue and green — can penetrate further (Fig. 8.5). The zone of the ocean which is sufficiently illuminated to permit photosynthesis is called the *euphotic* zone (the "well lit" zone, from Greek εὖ, *well* + φῶς, *light*). It generally extends to a depth of around 100 meters.

Figure 8.5 shows how deep light can be detected by some instrument located down there. In order that the light can return to the surface and be observed, the light has to be reflected by some material, in the swimming pool by the tiles, in the ocean by scattering particles or by the bottom (if it is not too deep). As the light received at the surface has passed twice the penetration depth through the water, *the sight depth*, the depth from where some light can be recorded at the surface, will be half of the distances shown in Fig. 8.5, i.e. around 100 meters for blue light. If we assume that the practical sight depth is due to reflection from the euphotic zone, the sight depth for the open ocean may be estimated at 50 meters. This seems reasonable as the sight depth reported for distilled water is 80 meters.

Water in shallow coastal areas tends to contain a greater amount of particles that scatter or absorb light wavelengths differently, which is why sea water close to shore may appear more green or brown in color. Note the difference in spectral distribution in the left and right diagrams of Fig. 8.5.

Note: Snow looks white when the light you receive has only been reflected from the surface and has not passed through the snow. When looking down into a deep hole in the snow, the light you receive will be bluish as all colors except blue have been absorbed by the snow surrounding the hole — the same situation as in

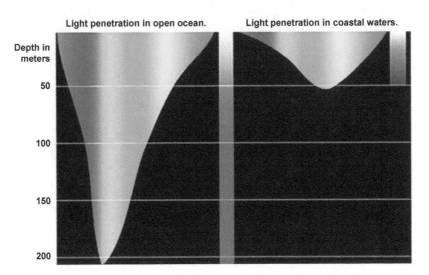

Fig. 8.5. Penetration depth in water. (Picture from Wikimedia.)

liquid water. Ice will also look bluish when looked through holes in it for the same reason. The color of *Glacier* ice is often shifted toward the green by fine particles in the ice.

Compare also with the color of clouds: Cirrus clouds high up in the atmosphere mainly consist of ice crystals. If the light you receive has not passed through the assembly of ice crystals but only been reflected, the cloud will look white — just like the snow field. Clouds in an approaching thunder storm will look dark blue if the light has passed through the enormous amount of water present in the cloud (in the form of water drops).

Crater Lake in Oregon, USA, is widely known for its intense blue color and spectacular view, Fig. 8.6. The lake is a caldera formed by the collapse of the volcano Mount Mazama. There are no rivers flowing into or out of the lake; the evaporation is compensated for by rain and snowfall. The water will in this way be replaced every 250 years. With a depth of 594 m it is the deepest lake in USA. The appearance of the lake varies from turquoise to deep navy blue depending on whether the sky is hazy or clear (reflecting the relative influence of sky and intrinsic scattering). The dependence of the color on the observation angle is quite noticeable in Fig. 8.6.

Fig. 8.6. Crater Lake. (Photo by the author.)

9

Electron Microscopic Studies of Snow Crystals

Low Temperature Scanning Electron Microscopy. In studying the finest details of snow crystals electron microscopy has turned out to be a very useful technique. At the Beltsville Agricultural Research Center in Maryland, one of the uses of the Low Temperature Scanning Electron Microscope (LT-SEM) is for the study of snow crystals. The pictures below have been taken at the Electron and Confocal Microscopy Laboratory at this research center. Information about the water content of the winter snow pack is, for instance, critical to the determination of the nation's water supply as well as protection from flooding. The following technique is then used.

Samples of snow and ice are collected from the face of the snow pack onto copper metal plates containing precooled methyl cellulose solution. Within fractions of a second these plates are plunged into a reservoir of liquid nitrogen which rapidly cools them to −196°C and attaches the pre-frozen material to the plates. Selected samples are transferred to the preparation chamber for sputter coating with platinum. This makes them electrically conductive and they are then placed on the precooled stage (−170°C) of the LT-SEM where they are imaged and photographed.

Fig. 9.1. Pictures with light microscope (*left*) and LT-SEM (*right*).

In the LT-SEM pictures it is possible to see details which are difficult to record by ordinary light microscopy. The difference is clearly demonstrated in Fig. 9.1 which shows light microscope and LT-SEM pictures of the same crystal (a depth hoar crystal from a snow pit in Wyoming). The external features in the LT-SEM picture are very sharp and the depth of field is very large. On the surface of the crystals there are often tiny dots, most likely consisting of small particles of ice which have not grown into the main crystal. The overall grainy surface is caused by the sputtering of platinum.

A Panorama of Snow Crystals.   A series of electron microscopic pictures are shown in Figs. 9.3a–h. When studying the crystals you should simultaneously take a look at the "Nakaya" diagram below where the dependence of crystal shape on temperature and humidity is shown (Fig. 9.2). The needles in Fig. 9.3d are often associated with heavy snow fall in the northeast United States. The magnification capability of LT-SEM is demonstrated by Figs. 9.3g–h.

Metamorphosis of Snow Crystals.   The snow on the ground gradually changes its shape, even at temperatures far below 0°C. The sharp branches on a dendritic crystal will then disappear, as shown in Fig. 9.4a. In the collection of snow crystals in Fig. 9.4b, the original shape is hardly recognizable.

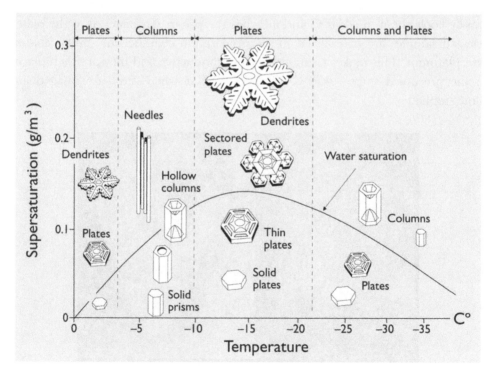

Fig. 9.2. Dependence of crystal shape on temperature and humidity.

Fig. 9.3a. Dendritic snow crystal.

Fig. 9.3b. Plate.
Note the double sheet structure.

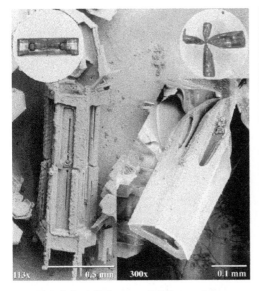

Fig. 9.3c. Columns of different form.

Fig. 9.3d. Needles.

Fig. 9.3e. Two plates joined by a column. A Japanese hand drum, Tsuzumi, to the left.

Fig. 9.3f. Three plates joined by three columns.

Fig. 9.3g. Details of a dendritic crystal.

Fig. 9.3h. Details of a dendritic arm.

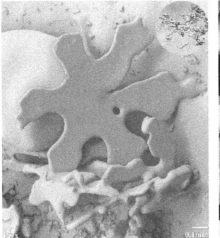

Fig. 9.4a. Aging of snow crystal.

Fig. 9.4b. Original shape no longer evident.

Firn and Depth Hoar Crystals.   Firn is formed under the pressure of overlying snow at the head of a glacier (Fig. 9.5a). This snow has undergone partial melting and recrystallization during a period of about one year (*firn* is a German name meaning "of last year").

In a snow pack, the snow close to the warm ground may evaporate and the water vapor desublimate on existing snow crystals above; these will then gradually disappear. *Depth hoar* crystals may thus be formed (Fig. 9.5b). These crystals ("sugar snow") are large crystals, up to 10 mm in diameter, occurring at the base of a snow pack. They bond poorly to each other, increasing the risk for avalanches.

Fig. 9.5a. Firn from South Cascade glacier.

Fig. 9.5b. Depth hoar crystal.

Surface Hoar, Rime, Glaze, Graupel and Hail.   The old English name *hoar* means "showing signs of old age." *Surface hoar* refers to the feathery, dendritic ice crystals that make trees and bushes look like white hair. It is formed when water vapor in the open air freezes, going directly from the gaseous state to the solid (*cf. depth hoar* above). *Rime* forms when fog or super-cooled liquid water droplets freeze on contact with cold surfaces; the water goes directly from the liquid to the solid state. Rime generally has a more icy solid appearance compared to surface hoar, but it is sometimes difficult to see the difference as rime may also be rather featherlike. Ships traveling through Arctic seas may accumulate large quantities of rime on the rigging. *Glaze* forms when super-cooled rain or drizzle freezes on exposed objects, such as telegraph wires and power cables. Glaze on the road surface can be extremely dangerous as the layer of ice (black ice) is typically thin, hard and transparent and the road appears to be wet rather than ice-covered.

Rime may also form on snow crystals as shown in the electron microscopy pictures in Figs. 9.6a–f.

Fig. 9.6a. Rime on a column.	**Fig. 9.6b.** Rime on plates.

Fig. 9.6c. Rime on a dendrites.	**Fig. 9.6d.** Rime on needles.

Fig. 9.6e. Graupel encasing a snow crystal (hidden).

Fig. 9.6f. Hailstone, 5 mm–15 cm in diameter.

As the riming process continues, the mass of frozen droplets obscures the original snow crystal and gives rise to a *Graupel* particle (Fig. 9.6e). (The German name *Graupel* probably comes from Serbo-Croat *krupa*.) The particle is also called *soft hail* or *snow pellet*. *Hail* is a more compact, spherical or irregular pellet, and the *hailstone* consists of a mixture of ice and snow (Fig. 9.6f).

Martian Snow.   On Mars the extreme low temperatures at the polar cap regions result in precipitation of carbon dioxide ("Martian snow") in addition to water-ice. Little is known about the structural features of the crystals. At the SEM laboratory, they have succeeded in producing and imaging CO_2 crystals. At high magnification, numerous 1 μm polyhedral crystals could be observed. The most common shapes of the crystals were octahedrons (Fig. 9.7). The results suggest that other gases, which precipitate at low temperatures and are of interest to interplanetary studies, may also be imaged using this technique.

Fig. 9.7. Crystals of carbon dioxide ("Martian snow").

10

Ice in Lakes and Glaciers

Basal Plane of Hexagonal Ice. Ordinary hexagonal ice, I_h, is the only type of ice occurring in nature. The crystal structure seen perpendicular to the hexagonal c-axis was shown earlier (Fig. 7.6). The structure seen along the hexagonal axis (*the basal plane*) is shown in Fig. 10.1. The water molecules form layers of hexagonal, puckered rings with chair conformation (shown more clearly to the right). The puckering (*cf.* Fig. 10.2) is due to the crystal symmetry of ice I_h, *P6₃/mmc*: In the structure there is not a *pure* sixfold axis but a sixfold *screw axis 6₃*, which means a combination of rotation of sixty degrees and translation ½ of the repeat distance in the c-direction.

Each water molecule forms three hydrogen bonds to its neighbors within the layer and one bond to one of the adjacent layers (this bond is directed perpendicular to the basal plane). One third of water molecules in the ring form hydrogen bonds to the layer above, one third to the layer below.

As each water molecule forms three hydrogen bonds to its neighbors within the layer and only one bond to an adjacent layer, the water molecules are much stronger bonded to its neighbors within the layer than to those in adjacent layers. The ice crystal is therefore mechanically very anisotropic. If stress is applied along the layers (perpendicular to the c-axis, *cf.* Fig. 10.2), the layers may glide on each other (Fig. 10.3). If stress is instead applied perpendicular to the basal plane, the ice crystal can still deform but the stress needed is 100 times larger than that required for basal gliding.

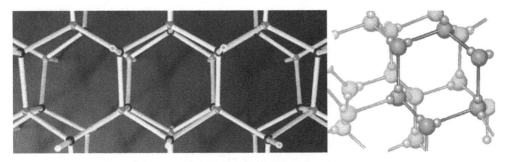

Fig. 10.1. The layers of puckered hexagonal rings with chair conformation.

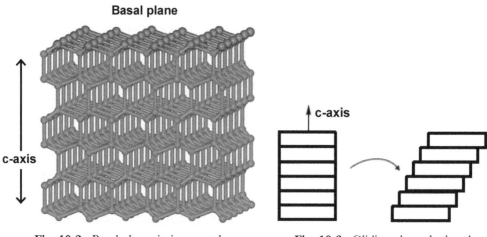

Fig. 10.2. Basal planes in ice crystal.
(© Martin Chaplin, Water Structure and Science. Permission granted.)

Fig. 10.3. Gliding along the basal planes.

Glacier Ice.   As mentioned in previous chapter, firn is formed at the head of a glacier after a period of about one year (Fig. 9.5a). After many years the ice crystals further down in the glacier will under the pressure transform into single crystals of different sizes. The size of the individual crystals depends on the depth. In the Greenland ice sheet, the crystals are between 1 mm and 10 cm in diameter. In the top layers, the crystals are generally small, but with time the smallest crystals are "eaten up" by larger neighboring crystals. Close to bedrock, the crystals can grow very big as the geothermal heat released from the bedrock increases the growth rate of the crystals (*cf.* depth hoar, Fig. 9.5b). The ice crystals will be randomly oriented as the snowflakes have settled randomly (Fig. 10.4a) (this and the following pictures are received from the Centre for Ice and Climate at the Niels Bohr Institute, University of Copenhagen). Some crystals are oriented favorably for basal gliding and others are not. As the ice deforms, the individual crystals in the ice slowly change shape as the basal planes glide past each other. This causes the individual crystals to rotate. Generally, the *c*-axis of the crystals rotate towards an axis of compression and away from an axis of extension (Fig. 10.4b). The effect of this is that deep down in the ice sheet the crystals are no longer randomly oriented but have a preferred orientation, which depends on the flow history. Thus, the flow history of the ice can be found from investigation of the crystal orientation at different depths.

The crystal orientation is determined by studying thin slabs of the ice, approximately 0.5 mm thick. When this slab of ice is placed between two crossed polarization filters, the individual ice crystals can be seen. The color of the single crystal depends on its orientation and the thickness of the slab. Samples around 0.5 mm result in red to blue colors. Crystals with the *c*-axis oriented parallel to one of the polarizers are black. In Fig. 10.5a are shown thin sections of the crystal structure at a depth of a few hundred meters, and in Fig. 10.5b from the middle of the ice sheet. In the top of the ice sheet, the crystals have a random orientation. This is seen as the crystals of the thin section (Fig. 10.5a) have many different colors. Deeper down, the deformation of the ice has resulted in preferred orientation of the crystals. Accordingly, most of the crystals in the thin section have similar colors — blue (Fig. 10.5b).

Lake and Sea Ice. Ice crystals as well as snow crystals appear in a countless large number of different shapes and with properties which can be quite different. Detailed knowledge of the properties of ice can be of great importance for many activities, and we may compare with the situation for the Sami people involved in reindeer herding. Lists of more than 50 different names for ice are found in the literature. A large number of scientific studies have been focused on the formation

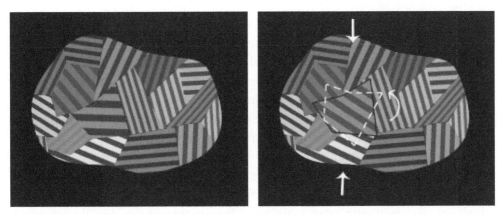

Fig. 10.4a. Ice crystals randomly oriented. **Fig. 10.4b.** Ice crystal reorienting under stress.

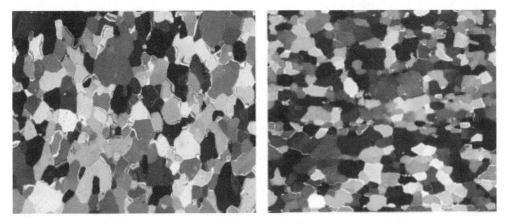

Fig. 10.5a. At a depth of a few hundred meters. **Fig. 10.5b.** At the middle of the ice sheet.

of ice, for instance on the orientation of the *c*-axis as a crystal grows under different conditions. Knowledge of the dominant crystal orientation is important in predicting albedo, mechanical properties and difference in stability (the albedo describes the reflecting power of the surface).

Let us first look at the situation when a snow crystal grows from the vapor phase close to 0°C. The morphology diagram shown earlier (Fig. 3.3) illustrates that the shape of the crystal is strongly dependent on the temperature: A temperature difference of only a few degrees in the region 0°C to −10°C may lead to a completely different shape. This reflects the fact that the growth rate of the different crystal faces is very temperature dependent. Close to 0°C, the crystal grows faster perpendicular to the *c*-axis; a few degrees lower the crystal grows faster along the *c*-axis. The fact that both plate-like and columnar snowflakes exist means that the ratio of prism and basal growth rates must change by a factor of 1000 under different conditions. A fundamental question is then: Why is there such a large difference in growth rate and why is the difference so strongly temperature dependent?

The environment is quite different when an ice crystal starts to grow in a lake. But is it possible that there is a similar strong temperature dependence of the growth rate of the different faces also when an ice crystal grows in water? And will the shape of the crystal accordingly depend on the temperature? To my knowledge no studies have been performed at several different temperatures in the range −1°C to −10°C under strictly controlled conditions in a laboratory. In many reports the focus has been laid on the orientation of the ice crystals under different conditions in Nature.

When fresh water freezes, a large number of ice crystals of different sizes will build up the ice sheet. These crystals often have a certain preferred orientation which has often been attributed to the growth anisotropy as just described. Crystals with a domination of both the *c*-axis vertical and horizontal have been reported. It seems to be unclear which conditions determine the development of the preferred orientation. The results have sometimes been divided in three groups, seeded and unseeded ice and slush ice.

Unseeded Ice.   According to some reports, large ice crystals with the *c*-axis vertical are formed when the meteorological conditions are suitable and no seed particles are available, as for example when no snow or ice fog falls onto the water surface and winds are light to allow formation of large crystals on the surface without being broken up by waves.

Seeded Ice.   If many seeding particles are available on the water surface, a large number of small ice crystals will be formed, with diameters around 0.05 mm to 20 mm. The seeding particles may come from snow falling onto the water or ice crystals which have been broken into small fragments by waves. Seeded ice tends form during cold, windy conditions. The first ice crystals that grow underneath the first layer may be both horizontally and vertically oriented but vertical crystals dominate further down according to some reports.

Slush Ice. Heavy snow may press down a thin ice layer so that the snow will be soaked with water. When this wet snow freezes, *slush ice* will be formed. In this ice, the crystals will be randomly oriented as the snowflakes have settled randomly. The slush ice is less mechanically stronger than the black ice.

It should be added that in the literature there are many conflicting reports. It is particularly difficult to determine the exact conditions when the ice crystals grow in Nature and the experimental data may be easily affected by systematic errors of one form or another. As the results are sometimes uncertain, it is difficult to draw definite conclusions about the conditions when the different types of ice crystals have grown and what factors have determined their orientation.

11

Gas Hydrates

History. Clathrate is the general name for a type of inclusion compound in which small molecules are trapped in a cage (from Greek *klaithra*, a bar, fence; from Latin *clathratus*, to furnish with a lattice). There are many different kinds of clathrates but the most common ones are the gas hydrates, where gases are enclosed in different kinds of cages formed by water molecules. Only these will be treated here.

In 1778, Joseph Priestley observed that some crystals were formed when bubbling SO_2 through 0°C water. About 30 years later, in 1811, Sir Humphry Davy discovered that a solid could be formed when a water solution of chlorine was cooled below 9°C. Davy has been considered to be the discoverer of gas hydrates, but Priestley was actually the first scientist to create gas hydrates in the laboratory.

It has been found that more than 100 different chemical species can combine with water. Typical substances forming solid gas hydrates include methane, ethane, propane, butane, nitrogen, oxygen, carbon dioxide, hydrogen sulfide, argon, krypton and xenon. The gas hydrates are "non-stoichiometric" which means that they have a composition where the proportions cannot be represented by integers. Most often, in such materials, a small percentage of atoms are missing or too many atoms are packed into an otherwise perfect structure. In the case of the gas hydrates, the reason for the non-stoichiometric composition is that some of the cages may not be filled. Depending on the type and size of the guest molecule, different structures will be formed. Three gas hydrates are described in this chapter.

During the first 100 years after Davy's discovery, these compounds were mainly of academic interest. However, in 1934, when the oil and gas industry in the United States was growing rapidly, it was recognized that the plugging of natural gas pipelines was not due to ice formation but to formation of gas hydrates. Natural gas leaving a gas reservoir is saturated with water, and when it expands into separators or at wellheads, the temperature drops and solid gas hydrates may be formed. Several techniques have been developed to prevent the formation of gas hydrates.

Hydrate formation as a step in methods to produce potable water from seawater received considerable attention in the 1960s and 1970s, but the processes have never been realized on an industrial scale. The effect of hydrate formation in food and biological systems has also been a subject of several investigations.

A major development occurred in the 1960s when it was realized that clathrates of natural gas exist in vast quantities in the earth's crust. In recent years, the large greenhouse effect of methane has been realized (20 times greater than that of carbon dioxide). An overwhelming number of papers have appeared, owing to the great interest in physics, chemistry, earth science, environmental sciences and engineering.

Structure of Gas Hydrates. The type of cage formed by the hydrogen-bonded water molecules depends on the size of the guest molecule. The pentagonal dodecahedron and tetradecahedron are the common cages in the gas hydrates.

The pentagonal dodecahedron is one of the five regular Platonic solids (the name originates from the ancient greek δώδεκα, dodeka, "twelve," and ἔδρα, hédra, "face" of a geometrical solid). It is a polyhedron with 12 pentagonal faces and can accommodate only one guest molecule. Such a cavity with a methane molecule inside is shown in Fig. 11.1. There are 20 water molecules in the polyhedron.

The tetradecahedron is a polyhedron with 14 faces, 12 pentagonal and two parallel hexagonal faces (Fig. 11.2). It is sometimes named tetrakaidecahedron (the greek "καί" means "and," i.e. a polyhedron with "four and ten faces"). In the picture, water molecules are located at each corner. There are 24 water molecules in the polyhedron. The volume of this cavity is slightly larger than the volume of the dodecahedron.

As pointed out in Chapter 12, *to form a closed structure, a polyhedron must contain 12 pentagons*. This rule is clearly fulfilled in both the polyhedra just described.

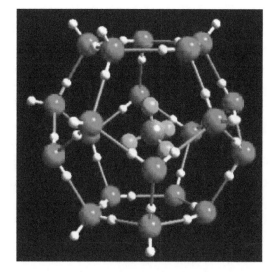

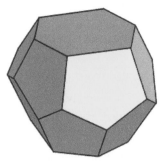

Fig. 11.1. Methane in pentagonal dodecahedron. (Picture: U.S. Geological Survey.)   **Fig. 11.2.** Tetradecahedron.

The interior angles in a regular pentagon are 108°, close to the HOH angle in water, 104.5°, and this is most likely one important reason why the dodecahedron and tetradecahedron are such common cages in the gas hydrates.

The gas hydrates usually form two different cubic crystal structures, Type I and Type II. A third, hexagonal, is seldom observed.

In the 12 Å unit cell of Type I, there are 46 water molecules, forming two small and six larger cages. The small cage is a pentagonal dodecahedron and the larger one a tetradecahedron. The ideal composition is $M.5\frac{3}{4}$ H_2O if all cages are

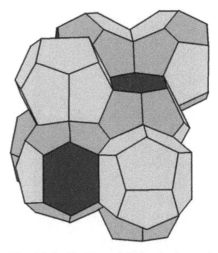

Fig. 11.3. Packing of dodecahedra and tetradecahedra in Type I gas hydrates.

occupied, and $M.7\frac{2}{3}$ H_2O if only the larger tetradecahedrahedra are filled. The packing of dodecahedra and tetradecahedra is illustrated in Fig. 11.3. It should be remarked that, because of their shape, the separate dodecahedra and tetradeca-hedra cannot form a compact arrangement.

The 17 Å unit cell of Type II contains 136 water molecules, forming 16 small and 8 larger cages. The small one is a pentagonal dodecahedron, the large one a hexadecahedron (a polyhedron with 16 faces). For simplicity the above polyhe-drons are sometimes called 12-hedrons, 14-hedrons and 16-hedrons, respectively.

Typical guests forming hydrates of Type I are methane and carbon dioxide; Type II hydrates are formed by gases like O_2 and N_2. It should be noted that a certain number of cages must be filled in order to make the hydrate stable. The empty cage alone is thermodynamically unstable.

The size of the guest molecule is the single most significant factor determining whether Type I or II is formed. In the crystal structure, each water molecule is hydrogen bonded to four neighbors, three within the cage and a fourth to a molecule in a surrounding cage. The water molecules in the hydrate cage are disordered and the reorientations of the water molecules result in approximately $(3/2)^N$ different configurations. This will lead to residual entropy at 0 K of the same magnitude as in Ice, I_h. The enclosed guest molecule interacts very weakly with the surrounding cage (van der Waals forces), and therefore in most cases it is relatively free to reorient. Whether there is a static or dynamic disorder will depend on the size and shape of the enclosed molecule, as well as the polar prop-erties of the molecule.

Fig. 11.4. Burning methane hydrate. (Photo: J. Ripmeester, NRC, Canada.)

Methane Hydrate. The hydrate cage encapsulating methane is illustrated in Fig. 11.1. It is possible to ignite lumps of the hydrate and it will then look like burning ice, as shown in Fig. 11.4. Methane hydrate is stable at 0°C in the oceans at depths below ~300 m. With increasing pressure the stability temperature rises to 24°C at 4 km depth.

Occurrence. Since the 1970s, methane hydrate had been found when drilling for oil in the form of layers or lumps in sand and gravel. On land it occurs only deep in northern permafrost regions, but in the sea it occurs on the slopes of the continental shelves and down to the deep ocean plains, both in the Atlantic and Pacific Ocean. In the cold water of the Polar Regions, it is found in rather shallow water, otherwise at depths towards 1,000 meter. It has even been found in deep lakes, like the Caspian Sea and Lake Baikal.

The potential reserve of gas in hydrate deposits has been estimated at over 10^{16} m^3, with about 97% offshore and 3% on land. According to early optimistic calculations, methane hydrate can produce twice as much energy as all fossil fuels — oil, gas and coal — taken together. But according to later more realistic estimations, the amount of energy is approximately comparable to that of natural gas. One complication is that methane hydrate that is mixed with sand and gravel is difficult to retrieve. India, China and Japan are spending considerable amounts of money to find technical solutions to make use of the available resources. For Japan, a domestic source of energy would be particularly valuable as most of its energy is imported. It is estimated that the amount of methane hydrate stored along the coasts of Japan would be sufficient to cover the needs for one century. A current field is located 50 kilometers outside the coast of Honshu, at a depth of 1,300 meter. Recent technical advances to retrieve the methane there appear quite promising.

Gas hydrates that naturally exist in the earth may pose severe problems to drilling operations: Casing damage, uncontrolled gas release, blowouts, fires and gas leakage outside the casing. Hydrates may also be formed in deep-water drilling. The pressure and temperature conditions are such that in most cases hydrates may be formed easily. In the past it has been reported that during drilling in deep-water wells, the equipment has become plugged and the formation of hydrates has caused great problems in subsequent operations.

Storage and Transportation of Natural Gas. It is believed that the natural gas that is today extracted in large quantities from gas fields to a large extent originates from hydrate deposits from which the gas has been released as the pressure in the reservoir is decreased (depressurization is one of the methods presently used technically to release the gas from hydrates). When 1 m^3 of gas hydrate is decomposed, 160 m^3 of gas (at 0°C and 1 atm. pressure) is released. It would clearly be a great advantage to store and transport the gas in the form of hydrate instead of in big tanks and tankers with thick walls (less pressure is needed in the case of hydrates). Could this be an alternative way of gas transport instead of in pipelines? Could a car be equipped with a small pressurized tank containing gas hydrate, from which gas is released by warming? To refuel the car would then also be a very quick matter. On the other hand, fossil fuel will hopefully not be needed in the future when electric cars become standard.

Effects Connected with the Release of Methane. The deposits of methane hydrate at the slopes of the continental shelves are mostly bound by sediment layers that act as a barrier against release of methane. The methane hydrate may remain stable in the sediments for long periods of time. However, the deposits far down may be heated by geothermal energy so much that the clathrate becomes unstable, resulting in release of methane. In this way very large resources of over pressurized gas may be formed, covered by hundreds of meters of clathrate-cemented sediments. Disruption of this protecting layer may result in an enormous release of the stored methane. Tectonic disturbances, volcanic eruptions, pressure drops due to lowering seawater levels, etc. are other factors that may also destabilize the clathrates.

There are several indications that big landslides with the potential of major gas release have occurred during earlier geological periods. At Storegga outside the Norwegian coast, such a big landslide occurred 8,000 years ago and 3.4 trillion cubic meters of material slid down the slope, resulting in a 15 m high tsunami. This is documented, as the tsunami wave washed organisms from the sea far up to the Scottish highland. The release of methane resulted in a large rise of the Earth's mean temperature for 10 years.

The importance of methane hydrate during older geological periods is hard to assess. It has been proposed that the massive extinction of marine life forms, such as the trilobites, about 250 million years ago at the boundary between the Permian and Triassic periods was caused by a sudden release of methane from methane hydrate on the sea floor. According to one hypothesis, the formation and decomposition of methane hydrate has played an active role in the fluctuations between ice ages and warmer periods.

During a warm period the water level of the oceans increases, the water pressure at the ocean floor increases and the methane there forms methane hydrate, decreasing the content of methane in the oceans. More methane can then be taken up from the atmosphere and the earth's temperature decreases (inversed greenhouse effect). More continental ice will form, the ocean water level decreases, the pressure at the bottom decreases and methane is released from the methane hydrate. A new, warm period, starts etc.

A similar role may have been played by carbon dioxide in the fluctuations between ice ages and warmer periods. There is a close connection between the surface temperature of the Earth and the atmospheric content of methane and carbon dioxide.

It has also been suggested that many of the sharp variations in carbon isotopes in the geological record may be related to methane release. The rapid diversification of species in the Early Cambrian could also have been driven by extreme environmental conditions caused by clathrate decomposition.

The Bermuda Triangle.   This is an area of the Atlantic Ocean where many ships sailing through it or planes flying over it are said to have disappeared without a trace. According to these stories more than 1,000 ships and planes have disappeared in the triangle area over the past five centuries. Many more or less fanciful explanations of these disasters have been suggested. One of these suggestions is of interest in the context of methane hydrate: "Methane gas trapped under the sea floor has been released in connection with a landslide. This has resulted in a lower water density due to the gas bubbles and caused ships to sink." However, there seems to be no truth in all these stories and explanations. Studies of weather reports on the days the incidents took place show that the meteorological conditions were not unusual. Tropical cyclones or hurricanes are for instance quite common in this area and the number of incidents in the area is not significantly larger compared to other ocean areas.

Hydrates of Carbon Dioxide.   The first evidence for the existence of CO_2 hydrates probably dates back to 1882, when Wroblewski reported the formation of clathrate hydrate in a system of "carbonic acid" and water. Carbon dioxide and water forms a hydrate of structure Type I under appropriate pressure and temperature conditions. If all the hydrate cavities are filled, the chemical composition is $8\ CO_2 \cdot 46\ H_2O$ or $CO_2 \cdot 5.75\ H_2O$.

The hydrate is stable up to $+10°C$ if the pressure is greater than 4.5 MPa (45 bar). Accordingly, the hydrates can appear from a depth of 500 m in CO_2-rich seawater (with a water pressure of 50 bar). These hydrates sink towards the sea bottom where they can stay for a long time (if undisturbed). It has been proposed to replace the gas in naturally occurring hydrate fields by carbon dioxide. The possibility to sequestrate carbon dioxide by storing it as solid hydrate in deep sea water has been considered, but this is presently at an experimental stage, and possible negative impacts to marine life must be thoroughly investigated.

For possible effects connected with the release of carbon dioxide, cf. methane hydrate.

Chlorine Hydrate. This hydrate, popularly called "chlorine ice," is one of the most stable gas hydrates and forms bright yellow-colored crystals. It has the highest melting point and a relatively low dissociation pressure at room temperature. Chlorine hydrate can be formed as wet chlorine gas is cooled and this sometimes causes problems, e.g. clogging in the pipelines when producing chlorine by electrolysis. The determination of the composition has attracted attention for a long time because of its non-stoichiometric nature. The composition depends on the preparation method and conditions. The formula of the hydrate can be written as $(8-x)Cl_2 \cdot 46H_2O$, where x is the number of unfilled cages in the elementary cell. The cage in which chlorine is enclosed is analogous to the other gas hydrates (*cf.* methane hydrate). From crystallographic structure studies (by among others Linus Pauling), it has been concluded that the Cl_2 molecule does not have a definite orientation in the cage and is probably able to reorient. The crystal structure is isomorphic with the other Type I gas hydrates; for a discussion of the packing of the polyhedra, see methane hydrate.

The use of chlorine in the form of chlorine hydrate instead of in gas form is technically of interest. For example, in the treatment of cellulose, chlorine hydrate has proved to be a safe and effective bleaching agent. It has given the best quality end product in the pulp manufacturing process compared to the use of chlorine either in the gaseous or solution form. Instead of transporting chlorine gas in tank cars or other thick-walled containers, chlorine can be transported as solid hydrate and liberated with warm water at the site of use. As pointed out above, employing hydrate formation as a means for storage and transportation has been has been considered also for other materials. There are two advantages: first, a much lower storage space is needed, and second, safety is improved. The challenge, however, is to develop economic methods for the production of the hydrates.

Encapsulated Water Molecules. A completely reversed type of clathrates has recently been made by researchers in Kyoto, Japan: Kei Kurotobi and Yasujiro Murata succeeded in enclosing a *single* water molecule in fullerene C_{60} (Fig. 11.5). In this fullerene, 60 carbon atoms form a truncated icosahedron, a polyhedron with 12

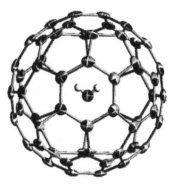

Fig. 11.5. C_{60} enclosing one water molecule.

pentagonal and 20 hexagonal faces (Fig. 11.7). Together with other researchers in Kyoto the technique has also been applied to enclose a single water molecule in C_{70} fullerene as well as a dimer of water in C_{60} and C_{70} fullerenes. These achievements open unique possibilities to study the intrinsic properties of water molecules!

Chemists have made several other attempts to encapsulate small molecules inside cage structures, with the aim to altering the properties of the parent structure. One of these examples is a helium atom trapped inside one molecule of dodecahedrane, $C_{20}H_{20}$ (*cf.* Fig. 12.3), which, according to the reports, is a quite stable structure. This substance is described as the world's smallest helium balloon.

The Truncated Icosahedron.   In geometry, truncation means to replace the corner of polyhedron with a plane (from latin "*truncatus*," cut off). The regular icosahedron is a polyhedron with 20 triangular faces (Fig. 11.6) (the name originates from ancient Greek εἴκοσι, *eíkosi*, twenty). The "truncated" icosahedron

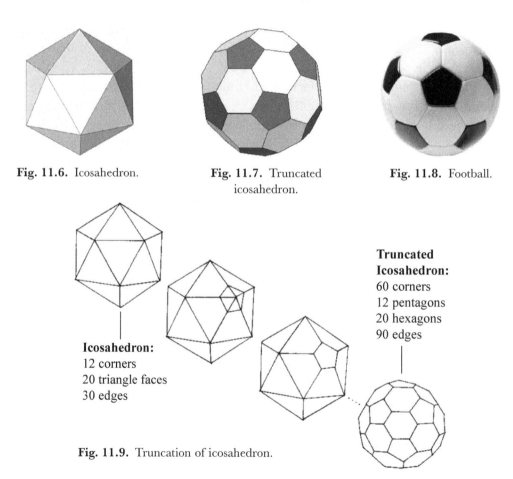

Fig. 11.6. Icosahedron.   **Fig. 11.7.** Truncated   **Fig. 11.8.** Football.
                                       icosahedron.

Icosahedron:
12 corners
20 triangle faces
30 edges

**Truncated
Icosahedron:**
60 corners
12 pentagons
20 hexagons
90 edges

Fig. 11.9. Truncation of icosahedron.

is a polyhedron with 12 pentagonal and 20 hexagonal faces (Fig. 11.7). It may be constructed from the icosahedron by cutting off one third of each edge and replacing it by a plane, as illustrated in Fig. 11.9. A football is shown in Fig. 11.8 for comparison.

12

Polyhedra Formed by Water, Carbon and Hydrocarbons

It is interesting to notice the similarity between the polyhedra formed by water in methane hydrate and by carbon in the fullerenes — in spite of the large difference in the entities forming the polyhedra — a single carbon atom versus a bent water molecule with three atoms!

The characteristic features of polyhedra formed with hexagons and pentagons are the following; this applies to the fullerenes as well as to the water clusters:

The polyhedra must have 12 pentagons in order to form a closed structure. And they obey the rule $a = 2(n + 10)$, where n is the number of hexagons and a is the number of carbon atoms or water molecules.

There are a large number of more or less stable fullerenes. A few examples:

$n = 0$: C_{20} (less stable due to strain)
$n = 20$: C_{60} (the famous Buckminster fullerene)
$n = 25$: C_{70}
$n = 26$: C_{72}
etc.

Another property of polyhedrons, the *Euler characteristic* (χ), is also of interest in this context: $\chi = V - E + F$, where V is the number of vertices, E the number of edges and F the number of faces. For any convex polyhedron we always have $\chi = 2$.

The structure of the smallest fullerene, C_{20}, is illustrated in Fig. 12.1. Note that the polyhedron, a dodecahedron, is composed of only 12 pentagons (as C_{20} is a convex polyhedron, $\chi = 20 - 30 + 12 = 2$). The oxygen framework in methane hydrate (Fig. 12.2) is exactly the same as in C_{20}! The interior angles in a regular pentagon are $108°$, close to the HOH angle in water, $104.5°$, and this is most likely one important reason why the dodecahedron is such a common cage in the gas hydrates (*cf.* Chapter 11).

There is an even closer similarity between the cluster in methane hydrate and in dodecahedrane $C_{20}H_{20}$ (Fig. 12.3). In both cases, hydrogen atoms are sticking out of the dodecahedron (however, hydrogen atoms are not sticking out from all of the vertices in methane hydrate).

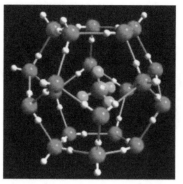
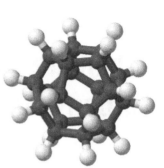

Fig. 12.1. Fullerene C_{20}.   Fig. 12.2. Methane hydrate.   Fig. 12.3. Dodecahedrane, $C_{20}H_{20}$.

Dodecahedrane is one of the *Platonic hydrocarbons*; the others are tetrahedrane, C_4H_4 and cubane (octahedrane), C_8H_8 (Figs. 12.4 and 12.5). None of them occur in nature, but once synthesized they are kinetically stable. Considering the similarity between the cluster formed by water and the cluster in dodecahedrane, would it possible to find clusters $(H_2O)_4$ and $(H_2O)_8$?

Can we find a water cluster composed of 60 water molecules like the structure of C_{60} (a truncated icosahedron, Figs. 12.6 and 12.7)? This is somewhat speculative, but the strain in the bonds in the hydrocarbon and water clusters is quite similar and the radial distribution functions from X-ray diffraction data actually give some support for clathrate-like structures in supercooled water. Would it be possible to form a *stable* water cluster by enclosing a large hydrophobic molecule inside a cage like in Fig. 12.8? Perhaps some hydrocarbon, which would correspond to a reversed situation as compared with that shown in Fig. 11.5, a water molecule encapsulated in a carbon cage!

In this context, the similarity in the bonding situation of diamond and ice should also be mentioned. In the crystal structures, the oxygen atoms in ice are located at the same positions as the carbon atoms in diamond. Note the tetrahedral environment in both structures (Figs. 12.9 and 12.10).

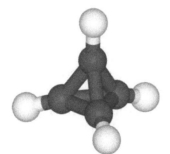
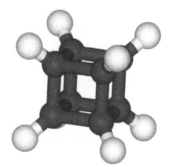

Fig. 12.4. Tetrahedrane, C_4H_4.   Fig. 12.5. Cubane, C_8H_8.

Fig. 12.6. Truncated icosahedron.

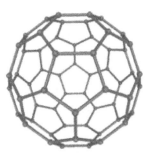

Fig. 12.7. Buckminsterfullerene, C_{60}.

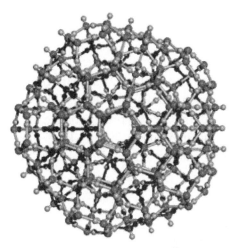

Fig. 12.8. Water cluster $(H_2O)_{60}$.

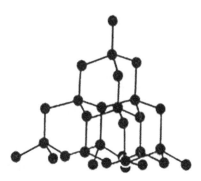

Fig. 12.9. Structure of diamond.

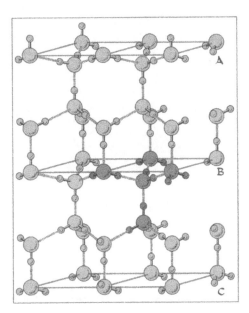

Fig. 12.10. Structure of ice I_h.

Regular polyhedra are formed by many other compounds, and also occur in the structures of living matter. Many viruses are either helical or icosahedral in shape. The icosahedron (Fig. 12.11) is a common polyhedron in spherical virus structures. It has been found that these viruses have a common icosahedral arrangement of their protein shells (the capsids). The icosahedral shape has been shown to be the most optimal way of forming a viral capsid for numerous reasons. It provides the virus with a very stable and almost spherical shape with a lot of room for its genetic material, the nucleic acid genome.

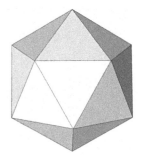

Fig. 12.11. Icosahedron.

An icosahedron has 20 triangular surfaces, which meet at 12 fivefold symmetry axes. A virus particle with 60 identical subunits (proteins) can, for instance, form a perfect icosahedron with three subunits on each triangular surface. The polio virus particle contains three times as many subunits, 3×60 proteins, blue, green and red, respectively (Fig. 12.12). It has a diameter of about 300 Å. Threefold as well as fivefold symmetry is noticeable in the arrangement of the subunits. The picture is taken from the VIPER data base.

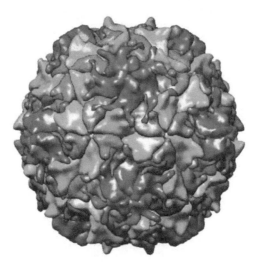

Fig. 12.12. Polio virus particle.

13
The Classical Elements of Nature

As the *classical elements* we usually consider the concepts in Ancient Greece of *earth, water, air, fire* and *aether*. They were proposed to explain the nature of all matter in terms of simpler substances, *elements*. Ancient cultures in Egypt, Babylonia, India, Tibet and Japan had similar lists, sometimes with local varieties of the names, e.g. "air" as "wind" and "aether" as "void" (empty). Water is also one of the five elements in Chinese philosophy, together with earth, fire, wood and metal. However, the elements in the Chinese Wu Xing system were understood as different types of *energy* in a state of constant interaction and flux with one another, rather than as different kinds of material.

The ancient Greek belief in five basic elements dates from pre-Socratic times. It persisted throughout the Middle Ages into the Renaissance and deeply influenced European thought and culture. The five elements are sometimes associated with the five platonic solids, as will be illustrated below. Water is an important component in the classical elements and it seems appropriate to provide some historical notes.

Water.   Thales (625–545 BC) is commonly named the first philosopher and was considered by Plato listed as one of the seven wise men. He lived in Miletus, an ancient Greek city on the western coast of Anatolia in Turkey. Aristotle considered him as the founder of the philosophy that asks what everything arises from. Thales asked himself: "What is unchangeable"; he looked for the *arche*, the origin of everything. His observation of the properties of water led him to an answer: *water* is the arche. Water can appear in different forms: solid, liquid and vapor. If everything consists of water also everything that happens can be explained if we know the laws and function of water. *Everything can be explained if we can explain water!* Even today we cannot explain all the properties of water so the conclusion is that we cannot explain everything!

At the time when Thales lived, there was a severe lack of water during the Olympic games and it happened that many spectators died from sunstroke. This is also how Thales met his fate at Olympia. Seven hundred years later, an aqueduct was constructed that led water from the mountains to Olympia, which helped the spectators to survive in the murdering heat.

Anaximandros (610–546 BC), a pupil of Thales, also tried to find some universal principle but assumed, like traditional religion, the existence of a cosmic order, a divine control of everything. He introduced the abstract *apeiron, the boundless infinite* as the origin of the universe, a concept probably influenced by Greek mythology.

Air.   Anaximenes (588–524 BC), the third philosopher of the Milesian school, was probably a pupil of Anaximandros but he was more attracted by Thales' idea that the arche must be something concrete in nature. Water only covers a limited part of the earth and such a substance appeared too limited to Anaximenes. Probably inspired by his teacher Anaximandros' somewhat diffuse and infinite arche, Anaximenes chooses the infinite atmosphere *air* as the arche. "Air is comparable to 'soul' and just as one's breath gives life, air gives life to all observable phenomena. When it is thinned it becomes fire, when it is condensed it becomes wind, then cloud, when still more condensed it becomes water, then earth, then stones. Everything else comes from these."

Fire.   Heraclitus of Ephesus (c. 535–c. 475 BC) considered *fire* to be the most fundamental of the elements and believed that fire gives rise to the other three. He regarded the soul as being a mixture of fire and water, with fire being the more noble part. He believed the goal of the soul is to get rid of water and become pure fire. Heraclitus is famous for his doctrine that change is the fundamental essence of the universe, as stated in the famous saying, "No man ever steps in the same river twice" (*panta rhei*, "everything flows").

Earth.   Empedocles (492–435 BC) was of the opinion that one arche is not sufficient as matter in the universe is so diversified, and introduced earth as a fourth element. What might have inspired Empedocles' ideas about matter? He lived in Acragas (the Roman Agrigentum), a town at the southern coast of Sicily. The volcano Etna is not far from Acragas and at the eruption, all the four elements are poured out: *fire* as glowing lava, *earth* in the form of gigantic stones, *water* in enormous downpours and hot *air* blowing out over the neighborhood, causing enormous disaster. According to one of the myths, Empedocles committed suicide by jumping into the glowing crater.

Empedocles is one of the most remarkable characters in the Greek history of culture. He considered himself as the master of the four elements and is sometimes compared to Goethe's Faust, the master over the elements as well as spirits and other supernatural forces. At the same time, he somewhere exclaims self-critically that "I realize that we cannot know anything."

According to Empedocles, visible matter is a mechanical mixture of the four arches, and it resembles the construction of a wall built up of four different bricks. Like Pythagoras, he believed in reincarnation: the souls can be transmigrated between humans, animals and even plants. "Wise people, who have learned the secret of life, are next to the divine and their souls, free from the cycle of reincarnations, are able to rest in happiness for eternity."

Aether.   In Greek mythology, aether was thought to be the pure essence that the gods breathed. It is also called *quintessence,* the material that fills the region of the universe above the terrestrial sphere. The concept of aether was used in several theories to explain various natural phenomena. In the late 19th century, physicists postulated that aether permeated all throughout space, providing a medium through which light could travel in a vacuum. Evidence for the presence of such a medium was not found in the famous Michelson–Morley experiment and the aether concept is today completely abandoned.

By the time of Plato (427–347 BC), the four Empedoclian elements were well established and it is not until in his cosmological dialogue Timaeus that aether is introduced as a fifth arche. The *Platonic solids* associated with the five elements are illustrated in Figs. 13.1–13.5. These polyhedra, the tetrahedron, cube (hexahedron), octahedron, dodecahedron and icosahedron are especially regular in the sense that all their faces are the same shape and the same number of faces meet at each vertex.

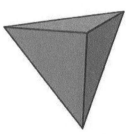 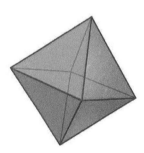

Fig. 13.1. Tetrahedron "Fire."   **Fig. 13.2.** Cube "Earth."   **Fig. 13.3.** Octahedron "Air."

Fig. 13.4. Dodecahedron "Universe."   **Fig. 13.5.** Icodahedron "Water."

14
Mysteries of Water

Escher's Waterfall and the Impossible Triangle. In folklore, we find a countless number of stories about various more or less imaginative mystical properties of water. We read about the wonderful healing power of water from holy springs as well as disasters caused by water, as for example, the Flood in the Bible. Some examples of the startling properties of water are presented below. It then seems appropriate to start this chapter with "Waterfall," one of the most well-known lithographs by the famous Dutch artist, M. C. Escher (1898–1972) (Fig. 14.1). To achieve the illusion of an eternally circulating water, he employed the same technique as in the Penrose triangle (Fig. 14.2). Two Penrose triangles are inserted in the waterfall to illustrate how it is used (Fig. 14.3) (in Fig 14.2, the triangle has been distorted to make it similar to those in the waterfall). The Penrose triangle also inspired Escher to make his other famous lithographs "Belvedere" and "Ascending and Descending."

Fig. 14.2. The Penrose triangle.

Fig. 14.1. M. C. Escher's "Waterfall."

Fig. 14.3. Construction.

The origin of the impossible triangles is worth commenting. In 1934, the Swedish artist Oscar Reutersvärd (1915–2002) made the first impossible triangle with cubes (Fig. 14.4). Since then he has created thousands of impossible figures and is today recognized as "the father of impossible figures." In 1954, the physicist Roger Penrose rediscovered this triangle, now commonly called the Penrose triangle.

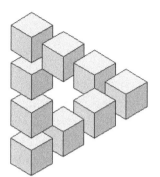

The Impossible Triangle Sculpture in Perth, Western Australia, is a nice illustration of how an illusion of a Penrose triangle can be achieved (Fig. 14.5). Seen from the correct angle, this sculpture seems to be a Penrose triangle.

Fig. 14.4. Triangle of cubes.

Memory of Water.   *Can water have a memory of its previous solutes, environment or processing?* Many hypotheses have been presented that water has the ability to retain a memory of substances previously dissolved in water even after a large number of serial dilutions. Some of the most famous cases are presented below.

Jacques Benveniste.   One of these ideas originates from the French immunologist Jacques Benveniste, at the time leader of a large INSERM laboratory in France (Fig. 14.6). In the late 1980s, he reported that certain biological effects seem to remain even at very high dilution of the biological substance (the action of very high dilutions of anti-IgE antibody on the degranulation of human basophils). Benveniste concluded that it was the *configuration* of the molecules in water that was biologically active as no molecules of the original substance could be present at this high degree of dilution. A journalist coined the term "water memory" for this hypothesis.

Benveniste's report was sent to *Nature* in 1988, but before publication, the journal needed confirmation by other independent laboratories. Researchers at

Fig. 14.5. Impossible triangle in Perth, Western Australia.
(Photo: Bjørn Christian Tørrissen.)

Fig. 14.6. Jacques Benveniste (1935–2004).

laboratories in four different countries confirmed Benveniste's results and the controversial paper was finally published after two years' delay — but with an unusual editorial reservation: "Readers of the article may share the incredulity of the many referees.... There is no physical basis for such an activity." After the publication, *Nature* therefore arranged for further independent investigations and another research team was set up. In cooperation with Benveniste's own team, the group could not replicate the original results. Although subsequent investigations gave the same negative results, Benveniste refused to redraw his article in *Nature*. He explained that the methods used in the later investigations were not the same as his own. His reputation was so much damaged that external sources of support were withdrawn. In 1997, he funded a new company to develop and commercialize applications of Digital Biology. According to one of his hypotheses, molecules can communicate with each other, exchanging information without being in physical contact and that at least some biological functions can be copied by certain energetic modes characteristic of a given molecule. In a paper from 1997, it was claimed that the water memory effect could be transmitted over phone lines, and in three papers from 1999 and 2000, it was claimed that it could be sent over the Internet.

Benveniste is the first person to be awarded more than one Ig Nobel Prize, the first in 1991, the second in 1998. The Ig Nobel Prizes are presented for "achievements that cannot or should not be reproduced." Nothing is said as to whether the achievement is good or bad, commendable or pernicious. Often the prizes first make you laugh and then make you think. The Ig Nobel Prizes are organized by the magazine *Annals of Improbable Research*, a magazine devoted to scientific humor.

Effect of Glassware. *Is the material used in the experiments important?* Glass is not indifferent to water and the process of dissolution of silicates from glassware has been much studied. It has been shown that even a relatively short exposure to water will lead to measureable amounts of silicates in the solution. It has been observed that ultrapure water can pick up 0.2 ppb Na^+ (10^{-7}%) from glassware in just one day. A famous case is shown by the story of polywater. The possible effect of the glassware in the preparation of homeopathic remedies is also discussed below.

Polywater. A lively debate took place between 1966 and 1971 concerning the possible existence of "super water." The excitement was the fact that in some experimental studies of water, a liquid had been observed with a maximum density of 1.4 g/cm^3, and which at $-30°C$ crystallized into a new previously unknown form of "ice." Some people were concerned about the risk that this super water could get out into nature and transform ordinary water with disastrous consequences. A large number of normally serious researchers threw themselves into the debate (hoping to receive a Nobel Prize?) and more or less imaginative proposals for the structure of the super water were published in scientific journals. After thorough investigations, it could be concluded that the liquid simply consisted of water that had been contaminated in the glassware used. Evidently, this water had a memory — that it had spent some time in a glass apparatus....

Homeopathy. The name originates from Greek ὁμοίως, *homoios*, "of the same kind" and πάθος *páthos*, "suffering." Homeopathy is a branch of alternative medicine, created in 1796 by Samuel Hahnemann (Fig. 14.7). It is based on his doctrine that *like cures like* ("*simila similibus curentur*"), the notion that a disease can be cured by a substance that produces similar symptoms in healthy people; and "law of minimum dose" — the notion that the *lower* the dose of the remedy, the *greater* its effectiveness ("potency").

Fig. 14.7. Samuel Hahnemann (1755–1843).

In the preparation of the remedy, a natural substance is diluted with distilled water or aqueous ethanol in a sequence of dilutions, each involving around 100-fold dilution. In each separate step, the container is vigorously shaken. The number of cycles is chosen so that the desired potency is obtained, up to 200 cycles; a larger number of cycles resulting in a higher potency. If one gram of the original substance is exposed to 200 cycles of dilution, $100^{-200} = 10^{-400}$ gram of the substance is expected to remain in the final product, *assuming that the dilution of the substance behaves normally* and *follows prevalent physical principles*. As there are only around 10^{23} molecules in one gram of the substance, it is clear that not one single molecule can be found in the final solution. Many different hypotheses have been presented by the supporters of homeopathy in attempts to explain how the final product can retain any power of the original substance. Some of the arguments are of the same type as those by Benveniste: The *configuration* of the water molecules is biologically active, even when no molecules of the original substance are present after such high degree of dilution, i.e. *water has a memory*. This hypothesis probably implies that the substance added has resulted in a particular arrangement of water molecules, and that this configuration is somehow retained in all the subsequent dilution steps (if the same glassware is used in all the dilution steps, the local situation at the glass walls is a factor to be considered). As glass is the preferred material of the container, vigorous shaking will certainly lead to small amounts of silicates in the solution (*cf.* above). From an analytical point of view, it would be interesting to know whether the homeopathic products contain traces of silicates, although certainly harmless.

Careful blind tests have shown that the effects of homeopathic drugs are placebo effects. This does not mean that no positive effects have been documented when patients have been given placebos. It is well known that a person's hopeful attitude and beliefs are very important to their physical well-being and that this is important for recovery from illness. According to current belief, the placebo effect is thus a psychological effect (*cf.* Fig. 14.8). A cornerstone in homeopathy is that the whole clinical picture is considered on an individual basis. Good care, interest and attention shown to the patient are important factors for a successful recovery.

Fig. 14.8. Snake oil man (immortal).

Masaru Emoto. Masaru Emoto (1943–2014) was a renowned Japanese author, researcher and entrepreneur (Fig. 14.9). One of his first discoveries was that all types of water do not form beautiful ice crystals. He found that polluted water (Fig. 14.10a) formed unpleasant deformed structures, while water from clean

Fig. 14.9. Masuru Emoto (1943–2014).

rivers, glaciers and from holy places formed beautiful crystals. Then he began to wonder if human beings can change the nature around us and tried many techniques to test this. He found that vibrations in music, written as well as spoken word, and even human thoughts can change the nature of water. He observed that beautiful ice crystals were always observed after speaking good words, playing good music or praying to the water. Disfigured crystals were observed when the opposite was true. Samples exposed to Bach's Air gave the expression that the crystal was dancing merrily; the word "angel" made the crystal burst forth in a multitude of flowers. The name Hitler made the crystal look like "I will kill you." He documented the changes in water by means of photographic techniques. Emoto published several volumes of a work entitled *Messages from Water*, which contains photographs of ice crystals and the accompanying experiments. The pictures below show ice crystals from water which has been exposed to different types of vibrations (Figs. 14.10b–h).

Emoto was personally invited to take part in the One Million Dollar Paranormal Challenge by James Randi in 2003, but he did not participate. The James Randi Educational Foundation offers to pay out one million U.S. dollars to anyone who

Fig. 14.10a. Bad water in Death Valley.
(Photo by the author.)

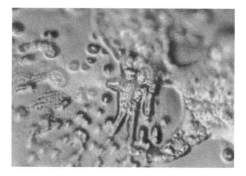

Fig. 14.10b. You make me sick.

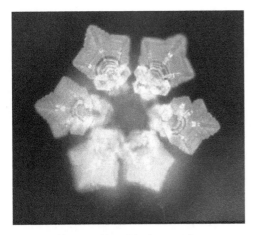

Fig. 14.10c. Playing music.
Imagine by John Lennon.

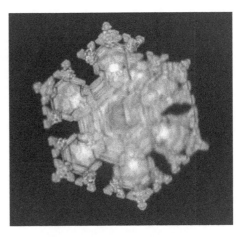

Fig. 14.10d. Shimanto river.

Fig. 14.10e. Showing
photo of dolphins.

Fig. 14.10f. Buddhist prayer
offered to the Fujiwara dam.

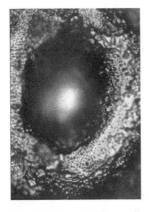

Fig. 14.10g. Talking evil.

Fig. 14.10h. Talking "eternal."

can demonstrate a supernatural or paranormal ability under agreed-upon scientific testing criteria. The challenge was first issued in 1964, and over a thousand people have applied to take it since then, but none has yet been successful.

15

The Mpemba Effect. Can Warm Water Freeze Faster Than Cold Water?

Since the time of Aristotle (384–322 BC) some scientists have claimed that hot water freezes faster than cold water. Aristotle writes in *Meteorologica*: "*The fact that water has previously been warmed contributes to its freezing quickly. Hence many people, when they want to cool hot water quickly, begin by putting it in the sun.*"

Aristotle supported an idea called *antiperistasis*, "the increase in the intensity of a quality as a result of being surrounded by its contrary. "Cold, on many occasions, increases a body's temperature, and dryness increases its moisture." The followers of Aristotle made extensive use of this principle. Even in the 1400s, scientists debated antiperistasis and could not decide whether human bodies were hotter in the winter than in the summer (as claimed by Aristotle).

Around 1461, the physicist Giovanni Marliani made some experiments with boiling and non-heated water and confirmed that hot water froze faster than cold. In the 16th century, similar observations were reported. Francis Bacon (1561–1626) has been called the "Father of Empiricism" and argued for *scientific* knowledge based only upon inductive and careful observation of events in nature. He writes "aqua parum tepida facilius conglacietur quam omnino frigida" (*slightly tepid water freezes more easily than that which is utterly cold*). His contemporary René Descartes (1596–1650) is also convinced: "*One can see by experience that water that has been kept on a fire for a long time freezes faster than other.*"

As modern theory of heat was developed, these early observations seem to have been forgotten by the scientific community, probably as they are contradictory to modern concepts. They did not become known by the modern scientific community until more than 2,000 years after Aristotle. The remarkable story of the rediscovery is as follows.

In 1963, a young student at the Magamba Secondary School in Tanzania, Erasto Mpemba, was making ice cream by mixing boiling milk with sugar. Instead of waiting until the mixture was cooled (as he was supposed to do), he placed it directly in the freezer. He was very surprised to find that his mixture froze into ice cream before that of the other students who followed the instructions. He asked his physics teacher for an explanation but got the answer that this was impossible. Erasto insisted that he was right but then his teacher taunted him, saying that this was not real physics, but "Mpemba physics." He had to believe his teacher but was not convinced. Later the same year, he met a friend who worked with ice cream in the town Tanga. Mpemba was told that he also put the hot mixture in the freezer and he found that other ice cream makers in Tanga had the same practice. Later, when Mpemba was in high school, he told his teacher about his experience with ice cream but got the same answer, that this is impossible. But when Mpemba tried the experiment with hot and cold water in the biology laboratory he again found that hot water froze faster.

Dr. Denis Osborne at the University College in Dar es Salaam had earlier visited Mpemba's high school to give a physics lecture. Erasto then went up to Dr. Osborne and asked him about the strange effect he had observed. Osborne had no explanation, but he was less sceptical than Mpemba's teacher and agreed to make some experiments later. He asked a young technician to test Mpemba's claim. The technician also found that the hot water froze first. "But we'll keep on repeating the experiment until we get the right result." However, repeated tests gave the same result, and in 1969, Mpemba and Osborne wrote up their results, lending Mpemba's name to the strange effect (E. B. Mpemba and D. G. Osborne, *Physics Education* **4**(3), 1969). Mpemba did not enter an ice cream research career, but instead moved on to study wildlife management. His discovery remained a mystery. Mpemba's discovery should remind us about the fact that many fundamental discoveries in the past are based on simple observations. Newton had an apple, Erasato Mpemba had ice cream.

A few interesting examples of the Mpemba effect in our daily life are presented in next chapter.

How Can We Explain the Mpemba Effect?   From the following experiment we would probably draw the conclusion that the Mpemba effect is impossible.

Start with two identical beakers with the same amount of water, one with water at 20°C, the other with water at 60°C. Apply exactly the same cooling process to both samples. Assume that it takes 10 minutes for the cold water to freeze. And assume that it takes 5 minutes for the warm water to cool down to 20°C. After that it should take another 10 minutes for this water to freeze. So it should take totally 5 minutes longer for the warm water to freeze!

However, even in strictly controlled experiments it has actually been found that warm water sometimes freezes faster. How is this possible? Many explanations have been proposed. Some of the factors to consider are:

1. *Surroundings*
- The material, size and shape of the beakers.
- The size and temperature of the freezer.
- How effectively/quickly the vapor is taken care of by the environment.
- The material on which the beaker stands.

2. *Convection*

• In the cooling process, convection currents will be developed and these will be more pronounced in the warm water. This may affect how quickly the water evaporates.

3. *Dissolved gases*

• Gases will be expelled from warm water and this may change the cooling properties.
• The freezing point may be slightly different for degassed water.

4. *Evaporation*

As the warm water cools down, it loses some water due to evaporation. When it has reached 20°C, a slightly smaller amount of water remains to be frozen. But it is difficult to calculate how much water has actually disappeared.

• How rapidly does the water evaporate (e.g. the influence of the surrounding air)?
• What is the heat capacity of the beaker? Etc.

5. *Supercooling*

• The tendency of supercooling may different for cold and previously warm water: the cold water contains more air bubbles and this may increase the tendency of supercooling (there will be a certain "dilution" of the water molecules and statistically less tendency of ice formation). The opposite effect has also been suggested but this seems less likely.

It is easy to realize that it is extremely difficult to arrange the experiment in such a way that the conditions are exactly the same for the two samples. It is, for example, virtually impossible to produce beakers which are identical and it would probably be best to make repeated experiments with the same beaker. But still: two experiments can never be exactly identical!

So, what conclusions should we draw?

• **Different conditions during the experiments may lead to different results.**
• **Under some circumstances hot water may freeze faster than cold water, but this does not imply that water does not obey the laws of thermodynamics.**
• **The fundamental laws of physics are still valid!**

The Mpemba effect is further discussed in Chapter 16.

16
Mpemba Effects in Our Daily Life

The early observations of the Mpemba effect were more or less forgotten by the scientific community. But even today the Mpemba effect is known as folklore among non-scientists in Canada and England, in the food industry and elsewhere. Are these observations just folklore or is there some truth in them? If these effects are actually real, such knowledge could be of considerable practical importance. A few examples are shown below.

Different attempts to explain the Mpemba effect have been listed in the previous chapter. In some of the cases discussed below, it is not easy to decide which of these explanations that is the most probable.

Snow Cannons.   In the snow cannon, water under high pressure is sprayed as small droplets into the open air at temperatures below 0°C. The droplets partly evaporate and the remaining droplets are cooled down by the heat of vaporization. Low humidity and cold wind will effectively stimulate the vaporization and formation of snow. One possibility to speed up the vaporization would evidently be to use *hot* water! In such a case, it would perhaps be possible to produce snow at temperatures higher than 0°C? To my knowledge this has not been tested, and it does not seem even recommendable today with all efforts to save energy! However, it would be interesting to make experiments with a snow cannon with both cold and hot water, and at air temperatures of 0°C and +5°C. This might become a nice illustration of the Mpemba effect on a large scale, *to show that warm water is more effective in making snow than cold water*! In this case, it is clearly vaporization that is the main origin of the Mpemba effect.

A simple way to demonstrate the Mpemba effect is to throw boiling water up into the air when it is cold outside. The difference between cold and hot water is demonstrated very clearly in the video by Louis Vincent at "https://youtu.be/uYBqZlWEOBc." Two pictures from the video are shown in Figs. 16.1–16.2.

Note that the water is very little spread out in this experiment (*cf.* left picture), compared to the droplets from a snow cannon. And it is therefore surprising that *all* the water has time to freeze, in a few seconds (right picture)! The cold water does not freeze at all! Why does the hot water freeze so quickly and *completely*?

A simple calculation of the energies involved in the above processes illustrates the effectiveness of *vaporization*. The energies involved in the different cooling steps

Fig. 16.1. T_{air}-25°C, T_{water} = +20°C   **Fig. 16.2.** T_{air}-25°C, T_{water} = ~ +90°C

are the following:

- It takes 539 cal/g for water to evaporate (2260 kJ/kg).
- It takes 1 cal/g to cool water one degree (4.18 kJ/kg).
- It takes 80 cal/g for water to freeze (334 kJ/kg).

To evaporate one gram of hot water (at 100°C), 539 cal is thus needed. This energy is obtained by cooling other droplets: if three grams of droplets are cooled from 100°C to 0°C, 300 cal are liberated. On subsequent freezing, $3 \times 80 = 240$ cal are furthermore liberated, totally 540 cal. *So, in principle, vaporization of one gram hot water is enough to freeze three grams of hot water!* Such a situation is of course very much simplified. All the heat of vaporization will not be used to cool other water droplets, but is lost to the environment.

The above experiment of throwing hot water into the air is a simple and perhaps the most convincing demonstration that the Mpemba effect is indeed a real, physical phenomenon. *Hot water may, under certain circumstances, freeze faster than cold water. In the above experiments, vaporization of some hot water droplets is responsible for the freezing of other droplets.*

From the above, it seems likely that the recommendation that a *car should not be washed with hot water* is based on correct observations. In this case, faster vaporization of hot water is also the mechanism.

Ice Cubes.   Assume that you find in your party that you are out of ice for the drinks. What is the quickest way to make ice cubes? Use hot water in the ice cube tray of metal! It is claimed that not only does hot water freeze quicker, but the ice cubes will also be of better quality (more transparent). If you want to get very clear ice cubes, you should boil the water a few times to get rid of the air bubbles. The air bubbles will make the ice cubes opaque when they are trapped in the ice. If you want to get large and clear crystals, you should also let the water freeze very slowly. It should be remarked that clear (pure) ice crystals melt more slowly than impure cubes!

Ice Rinks. The optimum conditions to make good skating ice have been studied extensively. The following facts appear to be well established.

1. *The water should be free from minerals*

The water which is sprayed on the rink freezes from the bottom up. As most compounds (impurities) are not incorporated into the ice when it freezes (due to the characteristic of the ice structure, *cf.* p. 54), the minerals will remain in the solution until the water at the top freezes. There will accordingly be an accumulation of minerals at the ice surface. This ice will become softer as the water molecules cannot form a perfect ice structure. The smoothness of the ice will also be reduced, resulting in increased friction between the skate blade and the ice.

2. *The water should be free from dissolved gases*

Dissolved gases will have the same effect as minerals on the ice; water free of gases will produce harder and more transparent ice. There are different ways to get rid of gases; the common method is to use warm water. According to ice-makers, the ideal temperature should be between 60°C and 70°C.

However, there is also another effect of using warm water: as discussed above, warm water under certain circumstances freezes faster than cold water. What is the most likely reason for the effect in this case? Evaporation and the absence of dissolved gases? Gas bubbles would probably increase the tendency of super-cooling and thus slow down the freezing.

One way to get rid of dissolved gases is to expose the water to very strong rotational motion, vortex. It is found that *cold* water which has been treated this way can be used equally well to produce excellent ice, and with considerable energy saving. The ice is also formed faster, becomes harder and tougher. The vortex process technology is also used for lime scale treatment and can be used to remove minerals from ordinary tap water.

3. *Air temperature and humidity*

You might think that colder air temperature in the rink results in better ice. But this does not seem to be the case: good ice can be made in a rink where the air temperature is 15°C. A relative humidity of around 40% appears to be favorable.

Hot-Water Pipes Break on Freezing While Cold Ones Do Not!

Plumbers have confirmed that this is correct. They have for instance observed that when two pipes are placed next to each other, one with hot water and the other one with cold water, the hot-water pipe breaks on freezing but not the other one. This may be explained as follows.

When the cold water pipe is successively cooled, the walls are not effectively kept warm and ice can start to be formed on the inside. The ice cover on the inside gradually gets thicker and thicker, but no pressure is exerted on the pipe walls in this process as the liquid water that remains can all the time be pressed away until the whole tube is filled with ice.

The walls on the warm water pipe are kept warmer and ice will not formed in the same way. At some stage, an ice plug may be formed and all the water close by may freeze simultaneously causing the pipe to break.

17

Hydrogen Bonding

The Hydrogen Bond. The hydrogen bond apparently fascinates youngsters (Fig. 17.1). Some basic facts about the hydrogen bond are summarized in the following.

Fig. 17.1. Watercolor by Rolf Lidberg (1930–2005).
(© Trollrike/Sweden. Permission granted.)

A hydrogen atom which is covalently bonded to a strongly electronegative atom X can under certain conditions be further attracted to a second electronegative atom Y under the formation of a hydrogen bond X–H···Y. Accordingly, the hydrogen bond is formed mainly between atoms such as oxygen, nitrogen, fluorine and chlorine. The bond energy is usually quite small, 20–40 kJ/mol (5–10 kcal/mol).

How can we make sure that it is actually a hydrogen bond?
A hydrogen bond between X–H and Y exists when

(a) There is evidence of bonding X–H···Y and
(b) That there is evidence that this bonding specifically involves hydrogen in X–H.

Evidence (a) can be verified by comparing bond distances obtained from diffraction studies and by requiring, for example, that the H···Y distance is shorter than the sum of the van der Waals radii for H and Y.

In many crystal structures, one finds C–H bonds which are directed towards more or less electronegative atoms Y at short distances, indicating a certain interaction. Today, such C–H···Y contacts are also called hydrogen bonds, although they are mostly considerably weaker than traditional hydrogen bonds. C–H···O hydrogen bonds are very common in organic crystals and play an important role in biological processes: They are responsible for base paring specificity, interactions of nucleic acids with proteins and the structure of amino acids, for example.

Evidence (b) can be verified by vibrational spectroscopy, noting that weakening of the X–H bond occurs on hydrogen bond formation, and as a consequence a decrease in the X–H stretching frequency. This frequency shift is accompanied by a substantial increase in the integrated intensity of the infrared absorption band associated with this stretching. Similar comparisons can be made using proton chemical shifts obtained from NMR, observing that hydrogen bonding results in a down-field shift in the proton resonance.

Models for the Hydrogen-Bond Interaction. Many different models have been suggested to explain the various phenomena associated with hydrogen bonding. The first requirement on such a model is that it offers a qualitative explanation of hydrogen bonding, i.e. that it accounts for the lowering in energy when a hydrogen bond is formed. More quantitative models should also account for the specific phenomena associated with hydrogen bonding. In particular, differences in bond strengths between different types of hydrogen bonds should be explained.

Early models tried to describe hydrogen bonding in terms of fixed point charges assigned to the atoms of the individual monomers, *the simple electrostatic model*. The interaction energy was then computed for given relative positions. However, only crude qualitative conclusions can be drawn from such a simple model. This model can be improved to any desired level of accuracy by introducing further charges around the nuclei to represent the total electron density of the individual monomers. However, such a refined model soon becomes less useful from a practical point of view and loses the simplicity of the original model. Many extensions of the models have been suggested, but today the only reasonable quantitative approach is to use non-empirical quantum mechanical calculations.

Linearity. Before the introduction of neutron diffraction in the early 1950s it was generally assumed that almost all hydrogen bonds are linear. However, deviations from strict linearity are quite normal for hydrogen bonds in solids, as well as in macromolecules. The hydrogen bond in ice can be taken as an example. In ordinary ice (I_h) there is a perfect tetrahedral distribution of the oxygen atoms,

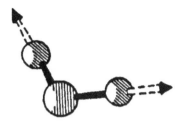

Fig. 17.2a. Two H-bonds donated.

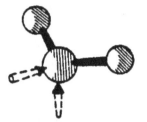

Fig. 17.2b. Two H-bonds accepted.

and the $O \cdots O \cdots O$ angle Θ is 109.5°. However, as the geometry of the water molecule in ice is practically the same as in the free molecule, 104.5°, the $O-H \cdots O$ hydrogen bond cannot be exactly linear.

The Role of the Lone-Pair Electrons on the Acceptor Atom. The hydrogen atoms in water and similar polar compounds are almost always engaged in hydrogen bonding. This fact has an important bearing on the arrangement of the hydrogen bonds, as can be illustrated for a molecule AH_2. Suppose that we wish to build up a three-dimensional structure containing only AH_2 molecules and require that all these molecules shall have the same bonding situation. Thus, if each molecule acts as a donor of two H-bonds, each molecule also must accept the two H-bonds (Figs. 17.2a–b).

In elementary discussions, one usually considers a lone pair as the acceptor of each of these H-bonds. With this picture, one would evidently need two lone pairs on each AH_2 molecule. However, one seldom finds the same number of "active" H-atoms as lone pairs. Water is a unique molecule in this respect and this is one important reason why water plays such an important role in all areas of chemistry and biology. *Water can act effectively as a donor as well as an acceptor.* In most other compounds, this ideal situation is not found; ammonia is a typical example. Here only one lone pair is available and one could then expect that only one of the three H-atoms will be engaged in H-bonding. The situation in solid ammonia is shown in Fig. 17.3.

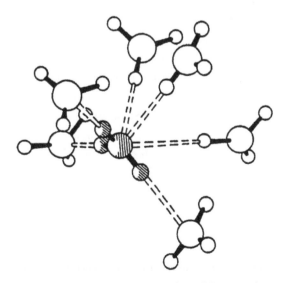

Fig. 17.3. Bonding situation in solid ammonia.

We notice that all three H-atoms participate in H-bonding. *The single lone pair has to accept no less than three H-bonds!* The simple picture of one lone pair per H-bond is evidently not relevant. From structural data, it can be concluded that the immediate acceptor of a hydrogen bond is a negative charge distribution, such as in the lone pair region, **but not specifically any individual lone pairs in the traditional sense.**

This is most likely true for all weak and moderately strong hydrogen bonds for which the major factor determining the hydrogen bond geometry is electrostatic interaction. In the case of very strong hydrogen bonds, covalent contributions may explain deviations from this picture.

The donor in a hydrogen bond, as, for example, an O–H group, attempts to direct itself towards a negatively charged region in the acceptor, but the finer details in the arrangement are often determined by other, geometrical factors, like the shape and size of the interacting molecules.

Another very important factor which must be considered is *cooperativity.* The oxygen atom in a water molecule will get a larger negative charge when it donates a hydrogen bond. The water molecule will then become a better acceptor for a hydrogen bond from a second water molecule.

In an analogous way, oxygen will get an increased negative charge when it receives a hydrogen bond. Hydrogen will then become more positively charged and the water molecule becomes a stronger donor of a second hydrogen bond. Such cooperative effects are important in all coupled systems of hydrogen bonds, not only between water molecules. They are probably very important in biological systems where a large number of hydrogen bonds may influence each other.

The Water Dimer. The water dimer has often been taken as a typical example of the directional influence of the lone pairs and many quantum mechanical *ab initio* calculations have been made to determine the conformation. The minimum energy occurs for a relative arrangement of the water molecules as illustrated in Fig. 17.4,

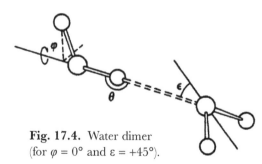

Fig. 17.4. Water dimer (for $\varphi = 0°$ and $\varepsilon = +45°$).

with $\varphi = 0$ and angle ε around 50°. (φ is the dihedral angle between the HOH plane of the donor molecule and the bisector plane of the acceptor molecule.) This might suggest a clear directional influence of one of the lone pairs of the acceptor molecule. However, as pointed out earlier, theoretical studies of the isolated water molecule show that there is a quite even distribution of the electron density in the lone pair region (*cf.* Figs. 7.3 and 7.4). A planar arrangement might seem equally probable ($\varepsilon = 0°$). How do you then explain the twisted (*trans*) arrangement? In our recent theoretical calculations, the minimum energy ΔE for different values of ε and φ was obtained:

$$\varphi = 0° \qquad \varepsilon = -45° \qquad \Delta E = -16.40 \text{ kJ/mol}$$
$$\varphi = 0° \qquad \varepsilon = 0° \qquad \Delta E = -19.41 \text{ kJ/mol}$$
$$\varphi = 0° \qquad \varepsilon = +45° \qquad \Delta E = -21.05 \text{ kJ/mol}$$
$$\varphi = 90° \qquad \varepsilon = 0° \qquad \Delta E = -18.20 \text{ kJ/mol}$$
$$\varphi = 90° \qquad \varepsilon = +45° \qquad \Delta E = -17.78 \text{ kJ/mol}$$

We notice that when ε is varied between $+45°$ and $-45°$ for a fixed φ value of $0°$, the interaction energy changes by 4.65 kJ/mol. A large part of this change can be attributed to a difference in the electrostatic repulsion between the non-bonded hydrogen atoms in the two water molecules (using a nominal value of $+0.4e$ for the partial charge on hydrogen). Furthermore, when ε is kept fixed at $+45°$ and φ is changed from $0°$ to $90°$, ΔE changes by ~ 3.3 kJ/mol, although the incoming H-bond directionality is not changing. The energy change corresponds well to the change in the repulsive non-bonded H–H interaction.

The above example may be taken as indications that there is no (or modest) directional influence of the charge in the lone-pair region of the acceptor water molecule and that the repulsion involving the hydrogen atoms not directly participating in the bond plays a significant role in determining the geometry of the dimer. In the *cis*-conformation ($\varepsilon = -45°$), the repulsion between the hydrogen atoms is the maximum and the binding energy correspondingly the smallest.

The Hydrated Proton

History. Hydration of the proton has long attracted considerable interest. It was early realized that the hydrogen ion H^+ cannot exist free in aqueous solution and the postulate that it occurs as a monohydrated species in aqueous solution is more than 100-years old. Goldschmidt and Udby at that time suggested that the complex ion "H20, H" is the carrier of the catalytic effect of the proton in esterification. In 1908, Hantzsch reported freezing-point lowering and conductivity measurements in solutions of sulfuric acid. The results suggested dissociation into HSO_4^- and H_3O^+ ions. The name "hydroxonium" or shorter "hydronium" for the complex is encountered here for the first time (today the name "oxonium" is normally used for H_3O^+). In 1912, Bagster and Steele found that anhydrous solutions of hydrogen bromide in liquid sulfur dioxide, which are practically nonconducting, become conducting on the addition of water.

The picture of a hydrated proton soon became more widespread. In 1922, Madelung wrote: "Das vermeintliche H-Ion ist also in wässriger Lösung zweifellos mindestens an eine Wassermolekel gebunden und entspricht dann als Hydroxonium OH$_3$ vollkommen dem Ammonium NH$_4$ mit dem einzigen Unterschiede, dass statt vier nur drei Wasserstoffatome an das Zentralatom gebunden sind."

From the high melting point of the monohydrate of perchloric acid (+50°C), it was early suggested that it is really a salt, oxonium perchlorate. If this assumption is correct, the monohydrate could be isostructural with ammonium perchlorate. In 1924, Volmer studied powders of the two compounds by X-ray diffraction. The two compounds gave almost identical diffraction patterns, a strong indication that the monohydrate of perchloric acid is indeed a salt, H_3O^+ ClO_4^-.

The abnormally high mobility of hydrogen in aqueous solutions led Hückel in 1928 to the suggestion that it was a result of some mechanism other than that usually used to describe ion mobility in solution. In 1936, Huggins proposed a mechanism where the doubly aquated proton, $H_5O_2^+$, could explain this abnormal behavior. Twenty years later, Wicke *et al.* found that the results of their calorimetric measurements of HCl solutions could be interpreted in terms of the complex $H_9O_4^+$ ($H_3O^+ \cdot 3H_2O$). In many other studies there was also found evidence that the four-hydrated form is predominant in solutions with sufficiently high water content.

It is very difficult to characterize the complexes of the hydrated proton in solution and even more difficult to determine their geometries. A large number of hydrates of strong acids have since been studied in the solid state by diffraction methods, and detailed information about bonding situation of the complexes in this state of aggregation is now available. In particular, neutron diffraction has provided detailed information about the geometry of the hydrated proton complexes.

The Hydrated Proton in Solids.   The crystal structure determinations of hydrates of strong acids have shown that the bonding situation of the proton can be described in terms of the basic complexes H_3O^+, $H_5O_2^+$, $H_7O_3^+$, and $H_9O_4^+$. Further water molecules are sometimes attached to form even larger complexes. The average geometry of these complexes is summarized in Fig. 17.5.

The oxonium ion, H_3O^+, occurs as an isolated pyramidal complex in compounds containing only one water molecule for each proton. The ideal symmetry is C_{3v}, but it is often distorted by the influence of its crystal environment.

In $H_5O_2^+$, the proton is located in the middle of the short hydrogen bond (2.44 Å) or slightly off-centered. This ion is characterized by a pyramidal bonding coordination around each end and it is a poor H-bond acceptor. The complexes $H_5O_2^+$ are formed in all cases in which exactly two water molecules are available for each proton. This ion is also found in many crystals containing more than two water molecules per proton. In such cases, the $H_5O_2^+$ ion is often bonded to the extra water molecules to form chains or layers of the type $\cdots H_5O_2^+\cdots$ $H_2O\cdots$ $H_5O_2^+\cdots$ The $H_5O_2^+\cdots H_2O$ bonds are here always considerably longer (≥ 0.17 Å) than the short bond within $H_5O_2^+$. This justifies the formulation $H_5O_2^+ \cdot nH_2O$ for these complexes.

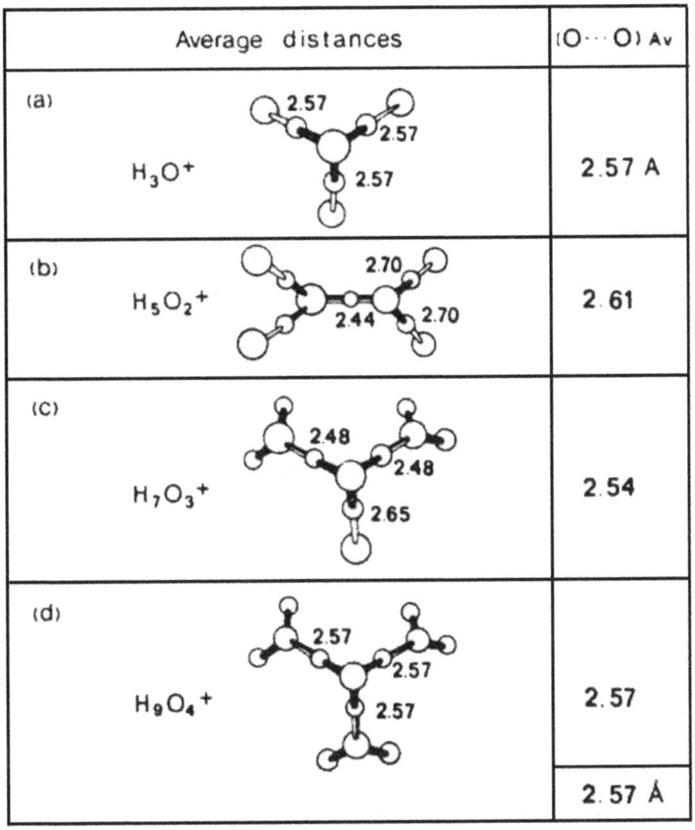

Average distances		(O···O) Av
(a) H_3O^+	2.57 / 2.57 / 2.57	2.57 Å
(b) $H_5O_2^+$	2.70 / 2.44 / 2.70	2.61
(c) $H_7O_3^+$	2.48 / 2.48 / 2.65	2.54
(d) $H_9O_4^+$	2.57 / 2.57 / 2.57	2.57
		2.57 Å

Fig. 17.5. Proton hydrates in solids.

The complexes $H_7O_3^+$ have been found both as isolated complexes and as complexes which are H-bonded to other water molecules. In the latter case, the $H_7O_3^+\cdots H_2O$ bond lengths are at least 0.22 Å longer than the H-bonds within the $H_7O_3^+$ ion. The internal structure can be very asymmetric and in some cases, it is a matter of definition whether the formulation should instead be $H_3O^+\cdot 2H_2O$ or $H_5O_2^+\cdot H_2O$.

In the complexes $H_9O_4^+$ of the Eigen type, the ion is in no case isolated but is H-bonded to water molecules or other hydrated proton complexes. The ion tends to be less well-defined than the other complexes. The $H_3O^+\cdots H_2O$ distance within the $H_9O_4^+$ complexes varies considerably (2.48–2.68 Å) and several alternative formulations may again be chosen: $H_3O^+\cdot 3H_2O$, $H_5O_2^+\cdot 2H_2O$ or $H_7O_3^+\cdot H_2O$.

From the above, it can be concluded that both the H_3O^+ ion and the centered or non-centered $H_5O_2^+$ ion are particularly well defined and occur frequently in the solid hydrates of the strong acids. The higher hydrates $H_7O_3^+$ and $H_9O_4^+$, although often well-defined with respect to their environment, are less well-defined in their internal structures and may be regarded as hydrates of H_3O^+ or $H_5O_2^+$.

Water in Biological Systems.   Water is of fundamental importance in biological systems since all living matter has evolved from and exists in an aqueous environment. There is a hundred times more water molecules in our bodies than the sum of all the other molecules put together. As hydrogen bonds are involved in most biological processes, increasing attention is today paid to the location of the water molecules around the biologically active molecules. It is realized that water actually plays an active role in the biological process and is not just an inactive partner. Due to its small size and flexibility in orientation, it is difficult to determine the details of the water structure and it is usually incompletely known. Not only the oxygen but also the hydrogen positions have to be accurately determined and neutron diffraction is the best alternative for large molecules. An increasing number of biological structures are determined with neutron diffraction as new facilities with stronger neutron flux have become available. However, often a major obstacle is still to get sufficiently large crystals.

Small Molecules.   In lower hydrates of carbohydrates, amino acids, peptides, nucleotides etc., the water molecules are not the primary factor determining the arrangement in the crystal structure. In non-ionic structures, the packing is mainly determined by the size and shape of the organic molecules, but in salt hydrates the water molecules may play an important role in completing the coordination shell around the cations. Molecules with a simple shape may pack effectively without the help of water, and accordingly form structures with low water content. Molecules with a more complicated shape, like disaccharides, nucleosides or oligopeptides form structures with a larger content of water.

Proteins.   They form two main groups, membrane proteins and water soluble proteins. In the first case, part of the environment of the proteins is non-polar and water does not generally participate in the interaction with the membranes. However, both on the inside and the outside surface, the membrane proteins interact with the solvent. In contrast, in a polar environment the hydration is very important for the structural, physical and biological properties. Also in the crystalline form, the macromolecules are strongly hydrated — between 20% and 90% of the total volume consists of solvent. In spite of the well-defined morphology, crystallized proteins resemble concentrated protein solutions. When a hydrated protein is cooled to liquid nitrogen temperature, the water which is directly bonded to the protein does not form an ice-like structure. In lysozyme, around 30% of the water is directly bonded to the protein, whereas the rest is less strongly bonded and more mobile. It is actually possible to replace this outer hydration shell with non-polar solvents without losing the catalytic activity. In contrast, the water molecules directly bonded to the active surface can usually not be replaced without losing the catalytic activity. The importance of the water molecules for the catalytic activity of lysozyme is illustrated by the fact that the catalytic activity starts almost at the same time as all the polar positions of the protein are completely covered by water molecules.

Surveys of the hydrogen bonding in proteins of known structure indicate that practically all backbone polar groups are involved in hydrogen bonds, either to intramolecular partners or to water. Unsatisfied backbone polar groups are energetically expensive and therefore unusual.

Nucleic Acids. In water solution, the double helix of DNA is completely hydrated, with around 20 water molecules per nucleotide, and the hydration appears to occur in two shells. Experiments indicate that 11–12 of the water molecules are directly bonded to DNA and this inner shell cannot be exchanged by ions; these water molecules do not form an ice-like structure on freezing. These water molecules function in certain cases as a direct link between proteins and DNA in the specific interaction between these biomolecules and the water molecules are apparently necessary components in the biological process. The second hydration shell may be penetrated by ions and on freezing, forms an ice-like structure. If salts or polar organic solvents are added to a water solution of DNA, the water activity is changed and water molecules are drawn from the outer hydration shell. The conformation of DNA is affected by the water activity and at high salt content, DNA will get other conformations than the biologically active B-form. In contrast, it appears that the conformation of RNA is not influenced if water is withdrawn from the outer hydration shell and RNA keeps its A-conformation under all conditions.

Water Transport in Trees. The transport system is quite complicated and in the following summary only the most important components are included. Some details strictly refer to conifer trees; the water transport is schematically illustrated in Fig. 17.6.

Water is transported from the roots to the leaves in a tube system made up of dead, water-filled and elongated cells with closed ends and thickened lignified walls (giving the mechanical strength of the stem). The tube system is called **xylem** (from Greek ξύλον, *xulon*, wood). The cells are arranged in cylinders parallel to the long axis of the stem but are shifted relative to one another. In conifers, the cells are called tracheids (from Greek τραχεία, *trachea*, airpipe and εἶδος, *eidos* – shape); the average length is up to 1 mm and the diameter 10–30 μm. The cells are of different lengths, but never long enough to reach all the way from the root to the leaves. The cells are therefore equipped with tiny pores to allow free passage of water from one cell to the other. The xylem system only passes water and mineral nutrients *upwards* through the stem.

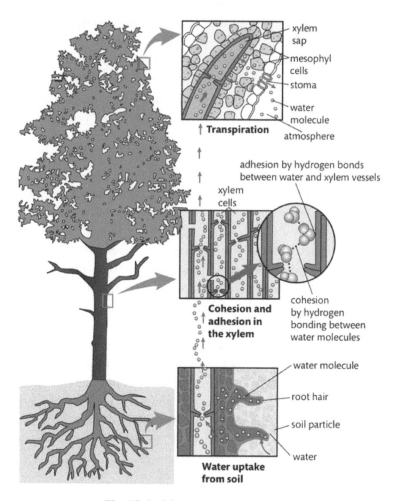

Fig. 17.6. Water transport in trees.

In vascular plants, ***phloem*** is the living tissue that carries organic nutrients (the photosynthetic products), in particular sucrose, to all parts of the plant where needed. In trees, the *phloem* is the innermost layer of the bark, hence the name, derived from the Greek word φλοιός, *phloios*, meaning "bark." The transport proceeds in both directions.

The xylem typically lies closer to the interior of the stem with phloem towards the exterior of the stem.

When the xylem arrives at the leaves, it branches into a fine network. The outermost offshoots extend all the way to the individual leaf cells. Water evaporates from the leaf surface through *stomata*, small adjustable pores which enable the uptake of carbon dioxide for photosynthesis and release of the oxygen formed. The evaporated water is replaced by sucking water from the xylem tubes by the capillary force from the leaves.

One fundamental question is now: What is driving the water from the root to the leaves? If water was transported through capillaries of radius 0.1 μm, the capillary force from the leaves would be enough to lift the water 100 meters, but the transport would be infinitely slow. The resistance (friction) from the cell walls would also be too large. The water transport requires mass flow and considerably thicker pipes. The system of tracheids with diameters of 10–30 μm accounts for this transport. The suction from the leaves is created by very fine capillaries on the rugged leaf cell surfaces; the capillary diameter is probably around 10^{-9} m. The total area of all these surfaces in a big tree can be of the order 10^4 m^2. A rough calculation shows that the capillary force from all these leaves may be sufficient to lift the water 100 meters in a tree with a radius of 1 m.

But there is one crucial factor that makes the suction from the leaves at all possible: *The water column in the xylem is kept unbroken, thanks to the hydrogen bonds between the water molecules.* There is a possibility that air is sucked into the column and cause cavitation, air bubbles that will break the water column. This is prevented in the following way: as mentioned above, the elongated cells in the long range transport system has walls that are equipped with fine pores so that water can move freely between the cells. A capillary force also acts locally in these pores, weaker than in the leaf cell surfaces but sufficient to bind water so hard that air is mostly prevented from being sucked in and initiating cavitation. Cavitated tubes can sometimes be refilled, but sooner or later all the tubes will suffer from cavitation. The tree, therefore, must make new tubes every year as compensation for those which have cavitated. The cavitated cells are filled with resinous material and polyphenols, and constitute the inner, darker part of the woody stem called heartwood. The outer, water conducting part of the stem is called sapwood.

Without hydrogen bonds, no water can be transported from the roots to the leaves!

18

Transformations of Our Earth by Water and Ice

The small and seemingly innocent water molecule will, when hydrogen-bonded to other water molecules, form a tremendously powerful tool to completely reshape our landscape.

According to the classical Chinese text *Tao Te Ching* ("the book of the way") "the highest excellence is like water. There is nothing in the world softer and weaker than water, and yet, when it comes to attacking things that are firm and strong there is nothing that can surpass it — because there is nothing that is so effective that it can replace water."

All over the world you will find a countless number of traces from the action of water and ice through the history of the Earth. Only a few examples, mainly from Sweden, are presented here.

Ice Ages. During the history of the Earth, it has been subjected to a countless number of large and small glaciations. Four major glaciations (ice ages) occurred during the Pleistocene, the geological epoch ranging from 2,600 million years ago to about 11,500 years ago. The last of these, called Weichsel, lasted for around 100,000 years. The name comes from the German name for the river Vistula in Poland, where the glacial maximum propagation reached. During this time, ice covered the entire Nordic region (except for a small part of Denmark), northern Russia, Greenland and much of North America.

During the warmer periods between the great ice ages, there have been periods of less cooling, even in recent times. Around 1500, northern Europe was suddenly hit by such a cooler period, which has been dubbed the Little Ice Age. Rivers and lakes froze during this period, which ended about 1850. Even a market could be held on the river Thames in England.

According to the "Snowball Earth" hypothesis, the whole earth has at least once been entirely or nearly entirely covered by ice. The proponents of this hypothesis argue that this best explains sedimentary deposits at tropical latitudes.

One hundred and forty thousand years ago, the ice that covered over the middle of Sweden was around 3.5 km thick. The surface of the earth was then pressed down so that much of Sweden was below sea level when all the ice had melted. The land gradually rose as the pressure of the melting ice decreased, and it is still rising, in the middle of Sweden, almost 1 cm per year. Since the latest glaciation the land there has risen 800 m. From Skuleberget at the World Heritage "Höga Kusten" (the "High Coast") there is a spectacular view over the surrounding landscape, as well as an adventurous trail (Figs. 18.1 and 18.2).

Fig. 18.1. View from Skuleberget at Höga Kusten.

Fig. 18.2. Natural crevice on Skuleberget ("Slåttedalsskrevan").

Fig. 18.3. Lapporten (Tjuonavagge, "The Lapponian Gate") in northern Sweden.

The tremendous force of the moving ice during the ice age may be illustrated by "Lapporten" in northern Sweden (Fig. 18.3). The originally V-shaped mountain edge was transformed by the moving ice into a "glacial trough" with U-shape. Many examples of U-valleys are found in other mountainous regions like the Alps, Himalaya and the Rocky Mountains. A classic glacial trough is in Glacier National Park in Montana, USA. The formation of a U-shaped valley can take anywhere between 10,000 and 100,000 years.

The effects of the violent water rushing from the melting ice are noticeable all over Sweden. Rocks of different sizes were deposited and formed the most common soil in Sweden, moraine (Figs. 18.4 and 18.5).

Fig. 18.4. Boulder-rich moraine. (Photo: Jan-Olov Svedlund.)   **Fig. 18.5.** Moraine with smaller rocks. (Photo: Jan-Olov Svedlund.)

Giant boulders occur in many places in northern Europe as well as in North America. These boulders may be up to tens of meters in section. The largest are difficult to distinguish from the bedrock, as they can be 30 meters high and have an area of one square kilometer. One example of a boulder transported by the ice is shown in Fig. 18.6. According to folklore, these boulders were thrown by giants towards churches as they could not support the sound of the church bells (Fig. 18.7).

The glacial striations shown in Fig. 18.8 are traces from the latest glacial period, commonly referred to as "The Ice Age." They are formed as the moving ice, laden with large quantities of rocks, scratches the bedrock.

Fig. 18.6. Giant boulder, transported by the ice.

Fig. 18.7. A giant's fling
Ill.: Robin Kuusela.

Fig. 18.8. Glacial striations at
Lake Huron, Ontario.
(Photo: Mary Sansverino.)

Fig. 18.9. Burial mounds in Old Uppsala.

Other traces are the eskers (ridges filled with small rounded rocks) (Fig. 18.9). The term *esker* is derived from the Irish word *eiscir*, which means "ridge or elevation". It is used particularly to describe long sinuous ridges, which are deposits of fluvio-glacial material. Eskers are frequently several kilometers long and are aligned parallel to former ice flow. The total length of the Uppsala and Stockholm eskers is estimated at 250 km and 60 km, respectively. Parts of the cities of Uppsala and Stockholm are built on these eskers, but the ridges are in many cases no longer visible due to excavations and erection of buildings. Eskers are often our biggest resources for drinking water. Where this water supply is not adequate, water from lakes or rivers may be purified by pumping it through the esker where it is slowly transported through the sand and gravel. The burial mounds in old Uppsala are part of the Uppsala esker (Fig. 18.9).

Giant's Kettles, Potholes.   Many large potholes were drilled in the bedrock during the ice ages when small rocks were caused to rotate by the water rushing out of the melting ice. After a very long time, a deep hole could this way be formed (Fig. 18.10). The largest in Scandinavia are 8–10 meters in diameter and 20–30 meters deep.

A large number of potholes can also be found along the coast, where the sea water has kept milling stones in perpetual rotation. In the *Annals of the Royal Swedish Academy of Sciences* from 1841, a story is told that illustrates old ideas on giant's kettles in the folklore. On the small island Öja (Landsort), southeast

Fig. 18.10. Pothole in a bedrock.

of Stockholm, a small pothole was found, today more than 2.5 meters above the sea level. It had a very small opening and a stone *larger* than the size of the opening was found inside (judging by the size of the stone, the diameter of the opening was probably smaller than 8 cm). The stone could accordingly not be taken out. Evidently some loose rock inside the pothole was caused to rotate by the sea water and a spherical stone was formed. It is a little hard to imagine that the sea water is able to form an effective vortex inside a hole with such a small opening. According to the superstitious inhabitants on the island the imprisoned stone was originally a malefactor who had been turned to a stone by the Gods, to be kept there until the day of judgement. The people on the island believed that if the stone was taken out, this would also become the end of the world. Probably in order to liberate the inhabitants from such superstition, Bishop Tingstadius in 1796 let break the pothole and the stone was taken out. The doom of the world did not come. The stone was donated to the "Upsala academi," the Royal Swedish Society of Sciences. The beautiful, perfectly spherical stone is shown in Fig. 18.11. The diameter is around 9 cm.

Fig. 18.11. The prisoner on the island Öja.

Beach Caves

Among the most spectacular along the coast are the tunnel shaped beach caves. Such a cave has an onion-shaped cross-section and is originally carved from a crack in the rock as illustrated by Fig. 18.12. The 13-meter long cave is located on the island Stora Kornö at the Swedish west coast.

Fig. 18.12. Beach cave on the island Stora Kornö.

Fig. 18.13. Beach cave at "Höga Kusten." (Photo: Johan Norrlin, SGU.)

Beach caves start to develop below the water surface, as far down as wave energy is active. These tunnel caves are of a special type that only seems to be documented in Scandinavia. More than 60 underground caves are known in Sweden. Caves mouths are all facing out to the sea. A large beach cave at "Höga Kusten" is shown in Fig. 18.13.

The Story of Döda Fallet (the "Dead Fall"). Around 1790, logging was a major industry in the heavily forested region of Jämtland, a province in the center of Sweden. The rivers were used as fast and relatively easy transportation of the timber to the coastal sawmills. However, the whitewater rapid Gedungsen (Storforsen) with a height of 35 meters was a major obstacle, as it damaged or destroyed much of the timber, forcing the use of land transportation past the waterfall. In 1793, Magnus Huss, a merchant in Sundsvall, for 100 crowns undertook to build a log flume past the much-feared Gedungsen. Early in 1796, the work on the canal was started. It was dug through unconsolidated glacial-outwash sand and gravel including an esker, starting downstream and gradually approaching the lake Ragundasjön. The spring flood of 1796 was unusually heavy, and lake water started to leak into the canal. The porous ground beneath the canal could not withstand the force of the water, which started to quickly erode deep into the esker. During the night between the 6th and 7th June, 1796, Ragundasjön drained completely in only four hours, triggering a 25-meter-high tsunami-like flood wave moving down the river, causing much destruction and establishing a new course of the river. Trees, boat houses, barns, dwelling houses, mills and cattle were swept away.

This is one of the biggest disasters in Sweden, but no one is believed to have been killed by the event. The washed-away soil and sediments redeposited as a delta in the Baltic Sea, creating new land on which the airport Midlanda was later built. The final judgment on the case (for loss of fishing) came in 1975, 179 years after the disaster.

Fig. 18.14. The once much-feared waterfall, Gedungsen.

At a rock barrier in the bottom of the former Ragundasjön, a new waterfall was formed at Hammarstrand, now turned into a hydroelectric power station. The dried waterfall Gedungsen is now called *Döda fallet* (the "Dead Fall"), and is a big tourist attraction (Fig. 18.14). The old lake Ragundasjön became fertile farmland.

> *"There is nothing in the world softer and weaker than water, and yet, when it comes to attacking things that are firm and strong there is nothing that can surpass it."*

> **The hydrogen-bonded water molecules have changed our world!**

19
The Rainbow

The Rainbow in Literature and Mythology.   The rainbow has challenged the fantasy of man since ancient times. The beautiful and inaccessible rainbow has become a natural symbol for human longing and dreams but has also been associated with fear of the unknown. It has given rise to a variety of tales and myths and has inspired authors, poets and painters as well as philosophers and scientists. The rainbow thus spans a bridge between the "two cultures," humanities and natural science.

In the Old Testament, the rainbow is a sign for the covenant between God and the earth:

> "I have set my bow in the cloud, and it shall be a sign of the covenant between me and the earth. When I bring clouds over the earth and the bow is seen in the clouds, I will remember my covenant that is between me and you and every living creature of all flesh. And the waters shall never again become a flood to destroy all flesh. When the bow is in the clouds, I will see it and remember the everlasting covenant between God and every living creature of all flesh that is on the earth" (Genesis **9**:13–16).

In the apocryphal book, Sirach the magnificent beauty of the rainbow is a proof of God's omnipotence.

> "See the rainbow and praise its Maker, so superbly beautiful in its splendor. Across the sky it forms a glorious arc drawn by the hands of the Most High (Sirach **43**:11–12).

In Greek mythology, *Iris* is the personification of the rainbow and messenger of the gods. She is also known as one of the goddesses of the sea and the sky. Iris links the gods to humanity. She travels with the speed of wind from one end of the world to the other, and into the depths of the sea and the underworld.

The ancient Greek philosophers did, however, leave the world of myths and made many significant contributions to our understanding of the phenomenon (*cf.* Aristotle below).

In Norse mythology, the rainbow is the bridge Bifrost, which goes from the Gods via Åsgård to the world of men, Midgård. The bridge is guarded by the god, Heimdal. Similar beliefs occur in other religions.

Goethe has sung the rainbow in many poems and there are many proofs of his keen observations. One poem deals, for instance, with an unusual rainbow, a "white" rainbow which is formed when the water drops are so small that the colors are mixed and the rainbow becomes fuzzy and white.

One even more unusual rainbow occurs in Schiller's drama, Wilhelm Tell. A group of farmers have met to plan for a rebellion during a night with strong moonshine. Suddenly, they notice a rainbow formed by the light from the moon — and even a second weaker one! It is very seldom that it is possible to see such weak rainbows with naked eyes.

The moon rainbow is also mentioned by Aristotle in his Meteorologica. He saw the phenomenon twice in 50 years.

Newton's analysis of the rainbow gave rise to a long and lively debate on the relation between the natural sciences and poetry. In his autobiography *The Immortal Dinner*, the English painter Benjamin Haydon gives a vibrant portrayal of the discussions during a dinner in 1817, when the famous essayist Charles Lamb and the poets William Wordsworth and John Keats were his guests. Lamb and Keats agree that Newton destroyed all the poetry of the rainbow by reducing it to the prismatic colors. The gaiety opens into "A toast to Newton, but down with the mathematics." Wordsworth participated in the gaiety and the famous toast but certainly with some inner reservation. Wordsworth had a completely opposite view: "Poetry has nothing to fear from the natural sciences, it incorporates the more narrow knowledge offered by the natural sciences. Poetry is the first and last in all knowledge." Newton had not reduced his feelings for the rainbow to something indifferent and trivial. This is demonstrated by his poem The Rainbow:

> *My heart leaps up when I behold*
> *A rainbow in the sky:*
> *So was it when my life began;*
> *So is it now I am a man;*
> *So be it when I shall grow old,*
> *Or let me die!*

The Physical Origin of the Rainbow.   The first attempt to explain the rainbow was probably made by Aristotle (384–322 BC). He described the phenomenon as a reflection of sunlight from clouds under a fixed angle. The eye receives rays which form a circular cone and Aristotle thus explained correctly the circular form of the rainbow.

After Aristotle, it took more than 1,600 years before any significant progress was made on the theory of the rainbow. In the 1300s a German monk, Teoderic of Freiburg, put forward the hypothesis that *each individual* raindrop is able to produce a rainbow. He thus rejected Aristotle's hypothesis that the rainbow is a collective reflection from the cloud as a whole. Teoderic even tested his hypothesis on a much magnified raindrop — a spherical glass bottle filled with water (Fig. 19.1):

light is sent from the right towards the bottle through a hole of the same size as the bottle. A faint rainbow will then be seen on the screen. He could follow the refraction and reflection of the light rays and concluded quite correctly the rainbow is formed by rays which have undergone reflection *inside* the water drop. He found that each color was spread in a direction which is typical for that particular color. To see the different colors, the eye accordingly had to be moved relative to the light-scattering water drop. From this Teoderic concluded that the different colors in the light, that reach the eye from the natural rainbow, originate from scattering by *different* water drops. The angles shown in (a) and (d) correspond to the deflection of red light. The illustrations (b) and (c) will be discussed later.

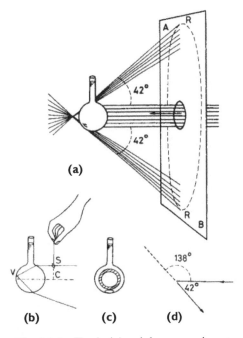

Fig. 19.1. Teoderic's rainbow experiment.

Teoderic's fundamental observations were little noticed and his results were largely unknown for 300 years. Independently, Descartes rediscovered these observations in 1637 and made further crucial contributions to our understanding of the rainbow phenomenon. Descartes explained the concentration of the scattered rays around a particular, fixed angle that gives rise to the rainbow and this effect came to be known "the rainbow effect." This is illustrated in Fig. 19.2: a series of parallel light rays that hit a water drop are deflected by refraction and reflection in the water drop. Ray 1 is redirected back and the deflection angle is 180°. The following rays 2–6 are deflected less and less with increasing distance from the center until a distance corresponding to Descartes' ray 7. After that the deflection increases again, rays 11–12. **Around the direction of ray 7 there is a strong concentration of rays.** The deflection of the rays is illustrated for one particular wavelength (a monochromatic beam). The beam path for Descartes' ray 7 is shown separately to the right. For a polychromatic beam, there will be a successive increase in the deflection angle, the so called rainbow angle, from red light (138°) to blue light (140°), as shown in Fig. 19.3. The angle between an incoming and outgoing red ray is 180° − 138° = 42°.

The collection of beams which contribute to the rainbow hit the spherical water drop almost tangentially within a narrow circular band with a diameter slightly less than the diameter of the water drop. In the setup illustrated in Fig. 19.1, it is accordingly possible to block these beams by hanging a narrow circular band of the same size in front of the glass bottle as shown in Fig. 19.1b–c. The rainbow will then disappear.

The brightest, the primary rainbow is formed by rays which are reflected *once* inside a water drop as shown in Figs. 19.2 and 19.3. Besides the primary rainbow, a secondary rainbow may quite often be seen. This is formed by rays which are reflected twice inside a water drop. The beam path in this case is shown in Fig. 19.4. Both the primary and secondary rainbows are shown in Fig. 19.5.

In the primary rainbow, the colors follow in the order red, orange, yellow, green, blue, indigo and violet. In the secondary rainbow, the order is reversed. Light which is reflected three times gives rise to a *tertiary* rainbow. For a possibility

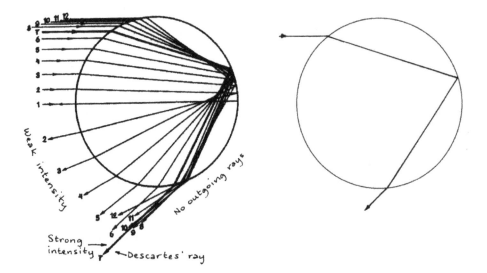

Fig. 19.2. Refraction and reflection by parallel rays in a water drop.

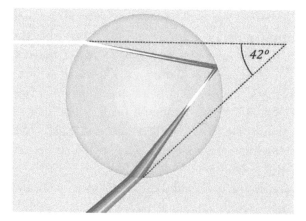

Fig. 19.3. Beam path in the primary rainbow.

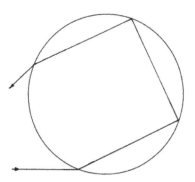

Fig. 19.4. Beam path in the secondary rainbow.

Fig. 19.5. Primary and secondary rainbows, with Alexander's dark band in between.

to see this, you have to look *towards* the sun; the angle between the direction to the sun and the tertiary rainbow is 40°. As this bow is very weak, it is in generally not seen unless the strong sunlight is damped, e.g. by a cloud. In principle, additional rainbows are formed by light which has been reflected more than three times. In the laboratory, rainbows of the 13th order have been seen by sending laser light towards a water drop — a sophisticated analogue to Teoderic's experiment 700 years ago.

In the angle range between the primary and secondary rainbows, no rays are reflected and this region will be somewhat darker than the surrounding sky. This region is called *Alexander's dark band* after the Greek philosopher Alexander from Afrodiasis, who first described this phenomenon around 200 AD. The band is faintly visible in Fig. 19.5.

The Geometry of the Rainbow. The rainbow is part of a circle, the center of which (the counterpoint of the sun) is located under the horizon (Fig. 19.6). The rays from the rainbow to the eye are located on the surface of a cone. The observer's eye is located at the tip of this cone. The axis of the cone is an extension of the straight line from the sun towards the eye. Half of the top angle of this cone is 42° for red light. The lower the sun is above the horizon, the greater part of the circle will be seen (*cf.* Fig. 19.5), until a half circle is formed when the sun lies at the horizon. If the sun stands higher than 42°, the rainbow disappears completely under the horizon. Only water drops which lie on the surface of the cone

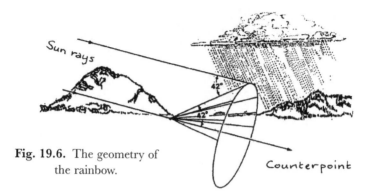

Fig. 19.6. The geometry of
the rainbow.

illustrated in the figure will contribute to the observed rainbow. The different colors that the observer registers originate from reflection in different water drops. The appearance of the rainbows is independent of the size of the water drops. Two persons standing side by side and looking at the rainbow observe in reality light which has been diffracted and reflected in different water drops. Everyone has his own personal rainbow.

Fig. 19.7. Rainbow with extra bows.

A very close look at the primary rainbow in Fig. 19.5 reveals some extra, very interesting features. Close to the violet region, in the upper part of the rainbow, a number of extra bows are faintly visible, normally as light red and green bands. The extra bows are more clearly seen in Fig. 19.7. These "supernumerary" bows cannot be explained by Descartes' model. They are due to inter- ference between coherent light rays around the rainbow angle for which the path length is slightly different. We may conclude that the interference effect which gives rise to these extra bows depends on the *size* of the water drops. The effect is more difficult to observe for large water drops and the extra bows can hardly be distinguished for drops larger than 1 mm. The size of the rain drops increases as they fall and the interference effect will accordingly be more noticeable in the upper part of the rainbow which originates from smaller rain drops.

There are many more interesting and subtle effects in the rainbows which cannot be treated here. As mentioned in the introduction, scattering of light from the moon may also give rise to rainbows, but the colors are generally so weak that they are perceived as colorless.

20

The Water Molecule is Unique

Water plays a unique role in chemistry. The special properties of the different forms of water — from ice and snow to liquid water — are due to hydrogen bonding $(O–H \cdots O)$ between the H_2O molecules. Of particular importance is the distinctive tendency of the water molecules to form hydrogen bonds with each other or with other polar molecules, and they can then operate both as donors and acceptors. The hydrogen bond is of fundamental importance in biological systems since all living matter has evolved from and exists in an aqueous environment. Hydrogen bonds are involved in most biological processes as little energy is needed in forming or breaking of these bonds. An important step in this context is proton transfer between donor and acceptor. Proton transfer in hydrogen bonds is one of the simplest chemical reactions and plays an important role also in many other fields of physics and chemistry. Without hydrogen bonds, no water can be transported from the roots to the leaves!

An extensive summary of recent research on water in its different forms, "Water Structure and Science" by Martin Chaplin is available in the Internet: http://www1.lsbu.ac.uk/water.

In this book, I have tried to illustrate the fantastic world of water in all its different forms. I encourage you to spread this message to all your friends — in the same way as the rivulets in the lava waterfalls Hraunfossar in Iceland. Here, the water from the glacier Langjökull flows underground around 20 km through the porous lava field Hallmundarhraun, until the rivulets are streaming into the Hvitá river, spreading over a distance of almost one thousand meters (Fig. 20.1).

Fig. 20.1. The lava waterfalls Hraunfossar.
(Photo: Ingeborg Breitfeld, Reykholt.)

Name Index

Subject Index

CPSIA information can be obtained
at www.ICGtesting.com
Printed in the USA
LVHW07*0311130318
569548LV00006BB/30/P